INTRODUCTION TO THIN FILM MATERIALS AND TECHNOLOGY

薄膜材料与技术导论

By Rujun Tang（汤如俊　编著）

图书在版编目(CIP)数据

薄膜材料与技术导论 = Introduction to Thin Film Materials and Technology:英文 / 汤如俊编著. —苏州: 苏州大学出版社, 2021.8
ISBN 978-7-5672-3579-3

Ⅰ.①薄… Ⅱ.①汤… Ⅲ.①薄膜 - 工程材料 - 英文 ②薄膜技术 - 英文 Ⅳ.①TB383②TB43

中国版本图书馆 CIP 数据核字(2021)第 133370 号

Introduction to Thin Film Materials and Technology

薄膜材料与技术导论

汤如俊 编著

责任编辑 沈 琴

苏州大学出版社出版发行
(地址: 苏州市十梓街 1 号 邮编: 215006)
广东虎彩云印刷有限公司印装
(地址：东莞市虎门镇黄村社区厚虎路20号C幢一楼 邮编：523898)

开本 787 mm×1 092 mm 1/16 印张 8.75 字数 192 千
2021 年 8 月第 1 版 2021 年 8 月第 1 次印刷
ISBN 978-7-5672-3579-3 定价: 35.00 元

图书若有印装错误,本社负责调换
苏州大学出版社营销部 电话: 0512-67481020
苏州大学出版社网址 http://www.sudapress.com
苏州大学出版社邮箱 sdcbs@suda.edu.cn

Preface

Thin film is a low-dimensional material widely used in many types of products. In recent years, the science behind this technology has developed into a comprehensive field that draws upon physics, chemistry, materials science, and vacuum and plasma technology. Thin film science and technology includes various methods for preparing thin films, such as physical vapor deposition and chemical vapor deposition techniques. It also includes the detection of the structure, composition and properties of films, as well as the application of thin films. Film research involves many advanced scientific concepts, including particle-beam interaction, material gasification/adsorption/nucleation, surfaces and interfaces of materials, amorphous and quasi-crystalline formation, impurities and micro defects, and so forth. The development of this field has promoted the development of condensed-matter physics, plasma physics, plasma chemistry, nano-materials, and so on.

These developments have given rise to a huge-scale film industry. Today's computers, communications equipment, control systems, detection instruments, and other technologies are made of microelectronic, optoelectronic, and micro-mechanical micro sensors and other devices. These devices are basically made of thin film materials. Therefore, the development of such materials is a limiting factor upon the development of electronic devices and even electronic equipment. Many countries regard thin film material and its preparation as strategic national interests. In recent years, the characteristic sizes of integrated circuits and other electronic devices have been reduced from micrometer-scale to nanometer-scale, offering higher integration and higher performance. This has also greatly promoted the development of excellent film materials, as well as new film-preparation technologies. Learning and understanding thin film technology has become a necessary requirement in high-tech fields like microelectronics, communications, and aerospace.

The purpose of this book is to provide an introductory understanding of the thin film material technologies for the bilingual undergraduate and technician to enter the new technical areas. In the first chapter of this book, the basic concepts and characteristics of thin film materials are introduced. The physical and structural-analysis methods of film materials are also briefly introduced. The second and third chapters discuss important background knowledge, particularly vacuum and plasma technology. The fourth and fifth chapters discuss the basic principle and physical-preparation method of gas-film-material and chemical-vapor deposition, including evaporation, sputtering, ion-beam, laser, and plasma-chemical vapor deposition techniques. In the sixth chapter, the growth physics of thin films are briefly discussed. The seventh chapter discusses methods for detecting the thicknesses and deposition rates of thin films. Finally, some recent advances of thin film materials and technology are introduced in the eighth chapter.

I heartily thank Professor Meifu Jiang for providing some of the materials in this book. This book can be used as a textbook for bilingual undergraduate courses in physics, materials science and engineering, electronic science and technology, and chemical engineering. It can also be referenced by scientific research and engineering technicians working in related fields. Due to editorial limitations and the rush of time, it is inevitable that this book would not be completely free of mistakes. I would like to invite readers, colleagues and experts to criticize and help improve it.

Rujun Tang

March 21, 2021

Contents

Chapter 1 Characteristics of Thin Films

From a macroscopic point of view, a *thin film* is a layer of material between two planes. Its thickness is much smaller than its other two dimensions. From a microscopic point of view, a thin film is a two-dimensional material consisting of atoms or clusters of atoms. But how thin is the scale to be considered for a film? There are no strict boundaries. *In general, thin film physics and processing studies are thin films between nanometers and micrometers in thickness.* Films with thicknesses larger than micrometer-scale are usually called thick films. Thin films are also man-made materials (usually made on a supporting substrate as shown in Figure 1.1); their structure and properties are closely related to their preparation methods and processing conditions. In this chapter, the basic microstructure and physical characteristics of thin films are briefly introduced.

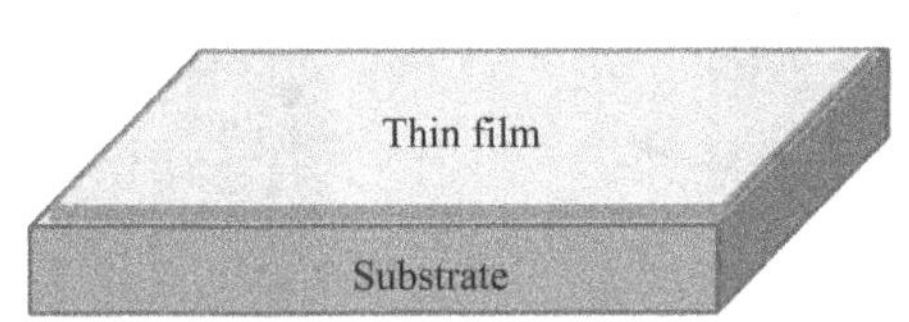

Figure 1.1 Schematic picture of a thin film material on the surface of supporting substrate

 1.1 Definition of films

In English, thin films have similar words including "coatings". The oldest instance of film preparation occurred in the Shang Dynasty of China, more than three thousand years ago, when our ancestors glazed porcelain. During the Han Dynasty, low-lead glazes were prepared using lead as a cosolvent. During Tang and Song Dynasties, Chinese glazing technology reached its peak. Such glaze coating serve not only as a beautiful decorative layer, but also increase the mechanical strength of ceramic materials. This makes it difficult to pollute and easy to clean. In modern times, the understanding of films began in the early nineteenth century, and it was found that solid films can be deposited via glow discharging. After the twentieth century, with the development of methods for preparing thin films such as electrolysis, chemical reactions, and vacuum evaporation, people began to study films from a scientific point of view. This has led to the rapid theoretical and practical development of thin film technology. Optical thin films were the first of many thin films to be studied. Many kinds of anti-reflection film, high-reflectance film, filters, and beam splitters have been precisely prepared and analyzed. These are also widely used in optical instruments, solar cells, building glass, and other fields as shown in Figure 1.2 (a), (b) and (c). Since the 1950s, the development of microelectronic devices has greatly promoted the progress of thin film technology. Thin film processes, including the deposition and etching of thin films, are the basis for the fabrication of integrated circuits, as shown in Figure 1.2 (d).

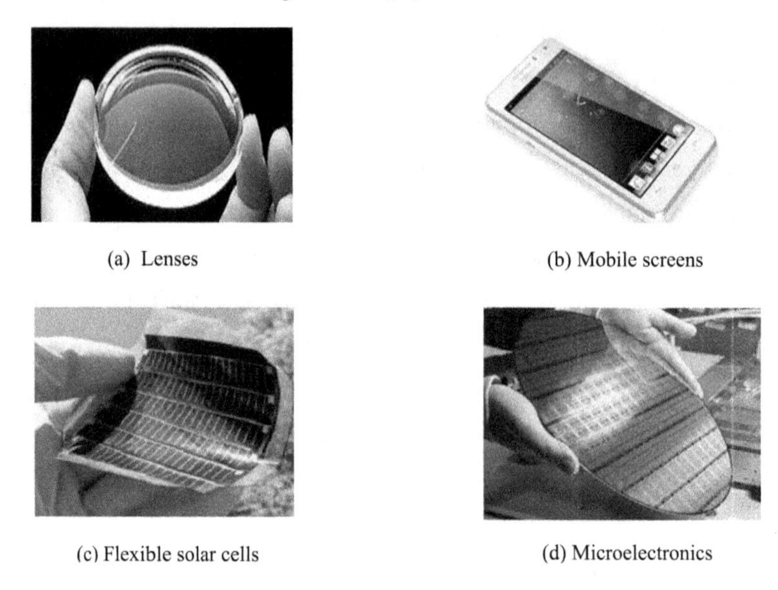

(a) Lenses

(b) Mobile screens

(c) Flexible solar cells

(d) Microelectronics

Figure 1.2 Exemplary thin film materials

The material that makes up a film may be gas (such as a gas film adsorbed on a solid surface) or liquid (such as an oil film attached to the surface of a liquid or solid); however, solid films are the most widely studied and applied. These can be inorganic materials, organic materials, or mixtures of the two. They can be either elementary or compound, and can consist of insulators, semiconductors, or metals. The structures may be single-crystalline, polycrystalline, amorphous, microcrystalline, or a superlattice.

What are the special structures and properties of thin films compared to the bulk materials? The possible differences will come from the surface effect, interfaces effects and small volume effects, which may result in different physical/chemical properties of the thin films from that of bulk materials. The crystalline defects in the thin films are sometimes also different. We will discuss the above special structures and properties of thin films one by one.

1.2 Surface effect

With the rapid development of thin films, it has been found that quantitative changes in material size will lead to qualitative changes in physical properties. When matter reaches a nanometer scale, even though the type of material has not changed, the physical and chemical properties of matter may have changed greatly. It is generally believed that the physical properties of a three-dimensional bulk material are independent of its volume, because its physical parameters within the bulk material are continuous. However, when the thickness of a material is reduced to a two-dimensional film state, some of its physical quantities will be interrupted at the material surface. *Its surface energy will be very different from its internal energy, leading to physical asymmetry at the surface.* Therefore, its electrical, magnetic, optical, thermal and chemical properties will differ from those in the bulk state. Here are some examples. Aluminum becomes a metal with excellent elasticity at the nanoscale; using a high-resolution electron microscope, we can observe that gold particles with diameters of 2 nm can form different shapes, including cubic octahedrons, decahedrons and icosahedrons; these nano-particles are not only different from aluminum's bulk state, but also different from its liquid state. Such particles are in a quasi-solid state with highly active surfaces.

1.2.1 Surface scattering

A thin film is a material with a large specific surface ratio (the ratio of its surface area to volume). We look at the effect of the film's surface upon the electron-transport phenomenon. Assuming that p ($0 \leqslant p \leqslant 1$) is the probability of electron and surface elastic collisions, $(1-p)$ is the non-elastic collision probability. The direction of the electric field along the surface of the film is x, the direction perpendicular to the film surface is z, and the distribution function is $f(z, v_x)$. The current density j_x along the x direction can be given by the following equation:

$$j_x = -2e\left(\frac{m}{h}\right)^3 \int_V v_x f \mathrm{d}V. \tag{1-1}$$

Here, e and m are electron charge and mass, respectively, and h is the Planck constant ($h = 6.62607015 \times 10^{-34}$ J · s).

According to the average value of $j_x(z)$ along the film-thickness direction, the conductivity of the film is obtained by the relation of the current density j and electric field:

$$\frac{\sigma}{\sigma_\infty}=1-\frac{3(1-p)L_\infty}{8d}. \tag{1-2}$$

Here, d is the film thickness, L_∞ and σ_∞ are the electronic mean free path and conductivity, respectively, in the bulk phase (film thickness is ∞). Obviously, as d decreases, the conductivity will significantly reduce and deviate from its bulk value. *The scattering effect upon the film surface will also affect its resistivity-temperature coefficient, Hall coefficient, thermoelectric coefficient, and so on.*

1.2.2 Melting-point reduction

When a solid ball with radius r dissolves, its interface energy between the liquid and solid will be ε, entropy change will be ΔS, mass density will be ρ, and dissolving heat will be L. When a solid of mass dm melts into a liquid, the changes in the surface area of the ball is dA and the thermodynamic-equilibrium relationship will be as follows:

$$Ldm-T_s\cdot\Delta S\cdot dm-\varepsilon dA=0. \tag{1-3}$$

For bulk materials, it will be:

$$Ldm-T_m\cdot\Delta S\cdot dm=0, \tag{1-4}$$

where T_s and T_m are the melting points of small-ball and bulk materials respectively. By substituting $\Delta S=\frac{L}{T_m},\frac{dA}{dm}=\frac{3}{\rho r}$ into Equation (1-3), we get

$$\frac{T_m-T_s}{T_m}=\frac{3\varepsilon}{\rho L_r}>0. \tag{1-5}$$

Thus, $T_s< T_m$; that is, the melting point of the small ball is lower than that of the bulk material. Moreover, T_s drops when the radius of the small ball r decreases. Taking lead (Pb) for example, when $r=10^{-7}$ cm is used, the melting point is $T_m-T_s=150$ K. That is to say, nano lead has a melting point 150 kelvins lower than bulk lead. The experimental results show that the *melting point of the thin film material is generally lower than that of the corresponding bulk material.*

1.3 Thin film structures and defects

The formation of thin films includes four processes: adsorption, surface diffusion, nucleation, and structure development. Hence, its final microstructure is closely related to its fabrication conditions, including gas pressure, gas-flow rate, power, substrate temperature, and so forth. Therefore, the microstructures and defects in thin films are different from and more complex than those in the bulk materials.

1.3.1 Film structures

Film structures can be divided into three types: crystal structures, surface structures and film/substrate interface structures.

1.3.1.1 Crystal structures

The microstructure of a film refers to its crystalline morphology, whether single-crystalline, polycrystalline, or amorphous. Schematics of these morphologies are shown in Figure 1.3:

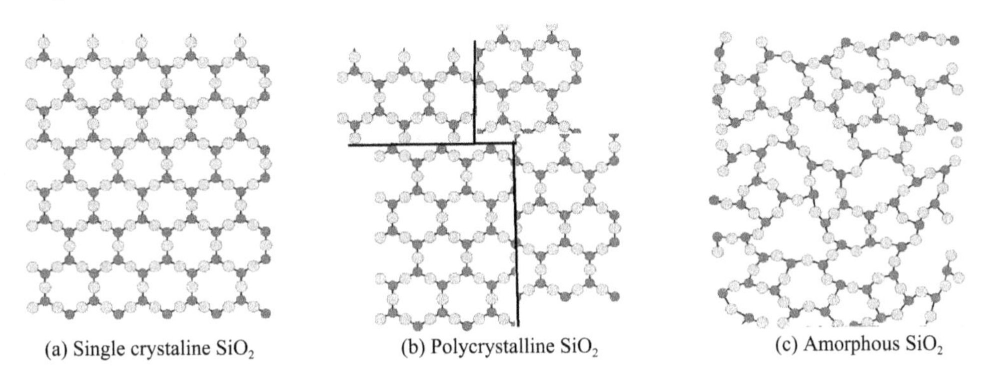

(a) Single crystaline SiO_2 (b) Polycrystalline SiO_2 (c) Amorphous SiO_2

Figure 1.3 Examples of crystalline morphology in SiO_2

1.3.1.1.1 Single-crystalline films

Under suitable conditions, such as on a single-crystalline substrate at a high temperature and a suitable deposition rate, a single-crystalline film can grow along the substrate's crystallographic axis; this is called *epitaxy*. Epitaxial growth is a common technique in the production of semiconductor devices and integrated circuits. To achieve epitaxial growth, three basic conditions must be satisfied: Firstly, the adatom has a high surface-diffusion rate, such that the substrate temperature and deposition rate will be very important. Secondly, the substrate is compatible with the film material. Assuming that the lattice constant of the substrate material is a and that of the film material is b, the lattice mismatch number m between the film and substrate is $m=(b-a)/a$. Smaller m values indicate a more similar lattice structure between the film and substrate, making it easier to achieve epitaxial growth. Thirdly, the surfaces of the substrates must be clean, smooth and chemically stable.

1.3.1.1.2 Polycrystalline thin films

In the process of film formation, many islands of small grains are formed. The films formed by the coalescence of these small grains are polycrystalline in morphology. Polycrystalline films are composed of grains of different sizes, and the boundary areas (planes) between grains are called grain boundaries. The grains of polycrystalline films can be arranged according to certain orientations to form different textures. For example, a fiber-structure film is a thin film with a preferred orientation; it is divided into a single-fiber structure and a double-fiber structure according to orientation, direction, and quantity. The former has only one preferred orientation for each grain, and the latter has two preferred orientations. The former is called a one-dimensional oriented film and the latter is called a two-dimensional oriented film.

The growth of ZnO piezoelectric films on glass substrates is a typical representation of fibrous-structured films. The film has excellent piezoelectric properties, which are basically related to the orientation of the crystallites. The preferred orientation of the *C* axis of each of the six crystallites in the film is perpendicular to the surface of the film or of the substrate.

In thin films, the preferred grain orientation can occur at all stages of film growth: initial nucleation, small-island coalescence, and the final stage. If the adatom has a higher diffusion rate on the substrate surface, the preferred orientation of the grain may occur at the initial stage of film formation. In the initial layer, the arrangement of atoms depends upon the substrate surface, substrate temperature, crystal structure, atomic radius, and so on. If the surface-diffusion rate of the adsorbed atom is small, the initial film will not produce the preferred orientation. When the film is thicker, the degree of the preferred orientation depends upon the substrate temperature, the incidence angle of the gas phase, and the deposition rate.

1.3.1.1.3 Amorphous thin films

Amorphous structure is sometimes referred to as glassy structure. In terms of the arrangement of atoms, it is a disorderly structure. For example, sulfides and halides films formed at low substrate temperatures tend to be amorphous structures. Some oxide films (such as TiO_2, Al_2O_3, etc.) may form amorphous films on room-temperature substrates. The diamond-like carbon films prepared by plasma chemical-vapor deposition are also amorphous at room temperature.

1.3.1.2 Surface structures

To minimize the total energy, the film should have the smallest surface area possible (i.e. it should be an ideal plane). In fact, this kind of film is not available. In the process of film formation, the vapor atoms are deposited on the substrate's surface and then diffuse laterally to occupy vacancies; this results in a reduction in surface area and a gradual reduction in surface energy. In addition, the adsorption and accumulation of atoms arrival firstly on the surface will affect the diffusion of the atoms arrival later, easily forming a "shadow".

The energy of lateral adatom diffusion on the surface is closely related to the substrate temperature. In general, when this temperature is high, the surface mobility of the adatom increases and condensation occurs preferentially at the surface concave or preferentially grows along some crystal planes. As the anisotropy and surface roughness increase the surface energy, the film tends to smooth the surface during the growth process. When the substrate temperature is low, because of the small-atom mobility, the surface will be relatively rough; when the area is larger, it will be easier to form a porous structure.

1.3.2 Defects in thin films

Imperfections in atoms in a film result in defects. For example, the growth of a film will lead to vacancies, dislocations, and adsorption of impurities, and also produce point defects, line defects, steps, grain boundaries, and so on. Generally speaking, the defect density of the film is often higher than that of the corresponding bulk material. The defects in the film will move and rearrange under the action of an external force. The film can be reprocessed by annealing and atmospheric treatment to change and control its structure and to improve the condition of the defect.

1.3.2.1 Point defects

For periodically ordered single-crystalline films, lattice vacancies, interstitial atoms and impurity atoms can destroy the periodicity of the lattice. If this happens within a linear range of one or more lattice constants, such defects are collectively referred to as *point defects*, as shown in Figure 1.4. Besides the interior point defects including bulk materials, there are many surface defects in films, which significantly influence on their structures and properties. These include surface vacancies, vacancies on the steps, surface-gap atoms, surface-alloy atoms, etc. The point-defect density in the film is much larger than that of bulk materials.

The formation of point defects may change during the film growth and even in the film-annealing process. For example, a vacancy may merge with other vacancies to form double vacancies, triple vacancies, or even voids. Point defects may also migrate to the surface of the film and disappear by being filled with other deposited atoms. Defects in the film, such as dislocations, can also cause surface-atomic changes.

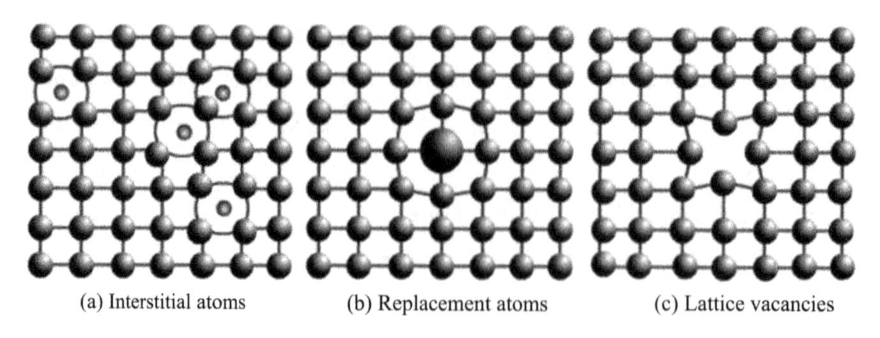

Figure 1.4 Schematic picture of point defects in a material

1.3.2.2 Line defects

When defects occur around a line inside a crystal, they are called *line defects*. Such line defects are mainly dislocations, including edge dislocations, screw dislocations, and mixed dislocations, as shown in Figure 1.5. In a crystal, a dislocation occurs on either side of an atomic plane (usually a dense surface), causing dislocations on both sides.

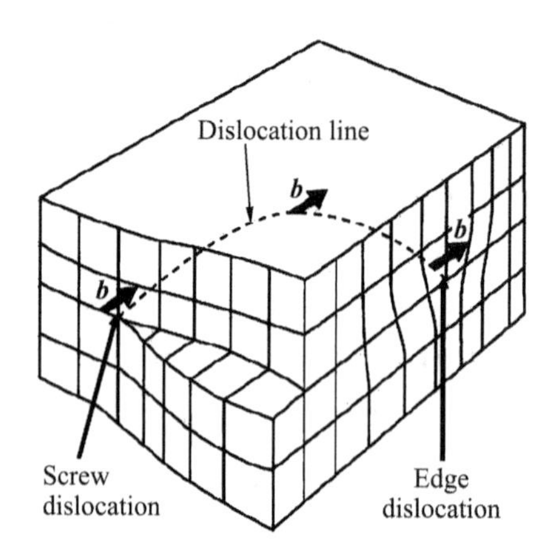

Figure 1.5 Schematic pictures of line defects in a material: edge dislocation and screw dislocation

In thin films, *dislocations tend to penetrate the surface of the film*. Part of dislocation body will pass through a surface and produce energy there, putting it in a pinning state. Unlike dislocations in the bulk materials, dislocations in the films are relatively difficult to move. They have relatively stable mechanics and thermodynamics, and are difficult to eliminate by annealing.

1.3.2.3 Planar defects

The *planar defects* in single-crystalline films are mainly twin boundaries and stacking faults. Crystals on either side of a twin boundary form mirror images. Stacking faults can be regarded as the adjacent two-twin interfaces (or twin boundary), as shown in Figure

1.6. In the process of film deposition, both twinning and stacking faults can be regarded as disorders of the stacking order of thin film atoms.

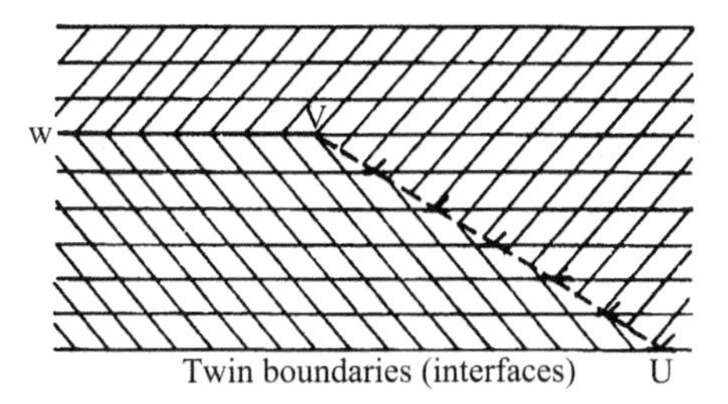

Figure 1.6 Schematic picture of planar defects in a material: twin boundaries (interfaces)

For polycrystalline films, inter-granular surface defects are also important. It is inevitable that the atoms at the boundaries connecting the two grains of different orientations will suffer from severe dislocation. When the misorientation of the two grain is small, the atoms on the grain boundary relax this distortion by means of local dislocations, which is the dislocation model of the small-angle grain boundary. Grain boundaries with two grain angles greater than 10 degrees are called large-angle grain boundaries. The crystalline (inter-granular) boundary has the same free energy as the interface of a general object; in general, the grain size of the polycrystalline material will change at higher temperature, and a large grain will gradually erode the small grain, which is manifested as the movement of the grain boundary. In this process, the grain boundary has a certain tension effect and also plays an important role in the transition of the solid phase. The newly formed solid phase begins to grow in many cases at the grain boundaries. Since atoms can easily diffuse along grain boundaries, exotic atoms can penetrate and distribute there. Internal impurities, atoms, or inclusions tend to be concentrated at the grain boundaries, giving them complex properties and producing various effects.

1.3.3 Anomalous structure and non-ideal stoichiometry

Most films are prepared in the non-equilibrium state, and their structures and phase diagrams do not necessarily agree. This is called an anomalous structure. This structure usually belongs to the metastable state, and it can be transformed into a stable structure under certain conditions.

An amorphous film is an abnormal structure. For example, the amorphous carbon film has a high hardness, good anti-friction performance, and low dielectric constant. It can be used as an insulating material. Graphite, however, is softer and has better electrical conductivity; it is often used as an electrode. The amorphous carbon film can be converted into graphite crystal after annealing. Amorphous structural materials have many electrical, magnetic, gas, and thermal properties. The amorphous materials have attracted widespread attention in the industry.

Again, the crystalline metal Bi does not exhibit superconductivity, but amorphous Bi films prepared by evaporation at 4 K have excellent superconductivity. A BN crystal has a hexagonal structure, but the BN film prepared by chemical-vapor deposition has a cubic structure, called cubic boron nitride (C-BN); its hardness is only inferior to that of diamond.

The ideal stoichiometric ratio of the compound is determined; however, the composition of the multi-component compound film tends to deviate from its ideal stoichiometry.

1.3.4 Film and substrate interfaces

Films are mostly attached to various substrates. Films and substrates form a composite system, and there is interaction between them. The adhesion and internal stress of thin films are the first topics to be studied in various applications. It is possible to study other properties of thin films only when they have good adhesion with the substrate. In addition, the structure of a film is greatly affected by the processing conditions during the manufacturing process, and certain amounts of stress are generated within the film. The thermal-expansion coefficient between the substrate and film materials also causes the film's yield stress. For example, if the bond is not firm, internal stress cannot be observed. Excessive internal stress will even peel off the film from the substrate. *Adhesion, diffusion, and internal stress are the inherent characteristics of thin films.* More discussions will be listed in Chapter 6.

1.4 Basic requirements to make a thin film

Thin films are man-made materials whose microstructure and properties are closely related to their preparation methods and processing conditions. How can we make a high quality thin film? *There are three basic elements to make a high quality thin film*, as indicated in Figure 1.7:

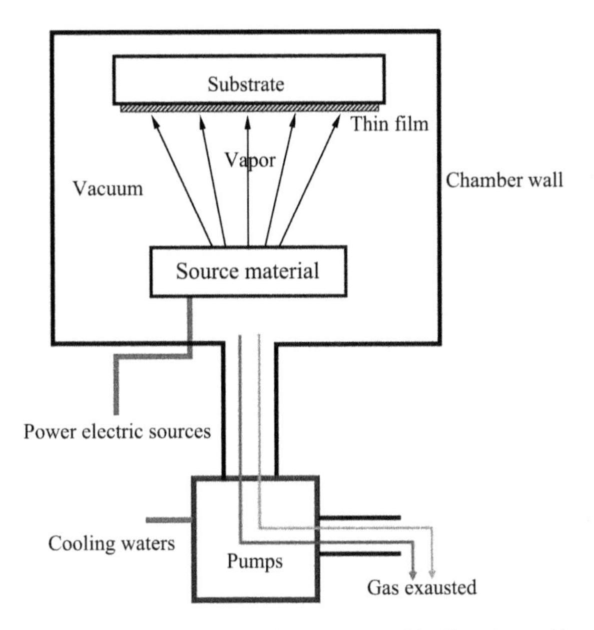

Figure 1.7 Schematic picture of the vacuum thin-film-deposition system

1.4.1 Chemical materials

Thin films are made from a combination of different chemical materials; as such, we need chemical raw materials to make thin films. Both these chemicals themselves and their reactants can be used as raw materials for thin film depositions.

1.4.2 Vacuum background

The quality of a thin film is very sensitive to lattice defects in the film. The dust, particles and chemically active gases in the atmosphere can reduce thin film quality. Thus, for a thin film deposition process, a high-background vacuum is a basic requirement.

1.4.3 High-energypower supply

In order to break the chemical raw materials into a vapor-like material, high energy is needed. For example, during the thermal-evaporation-deposition process, high temperature is needed; during the magnetron-sputtering-deposition process, a high-power electric field is needed.

More discussions of the above requirements will be listed in Chapters 2, 3, 4, 5 and 6 respectively.

Exercises

1. What's a thin film? Why are thin films so important?
2. Is a thin film an organic material? Is thin film a nanomaterial?
3. How are atoms bonded together to form thin film materials?
4. What are the differences between a grain and a grain boundary?
5. What kinds of defects can appear in a thin film? Please draw pictures to describe them.
6. What are the three basic requirements to make a thin film?

Chapter 2 Vacuum Technology Foundation

The deposition of thin film materials is usually carried out in a vacuum chamber. The detection of the physical properties of thin films also requires a vacuum environment. Therefore, vacuum technology is essential for film deposition and performance testing. To obtain a vacuum environment, various pumping systems are typically needed, as shown schematically in Figure 2.1.

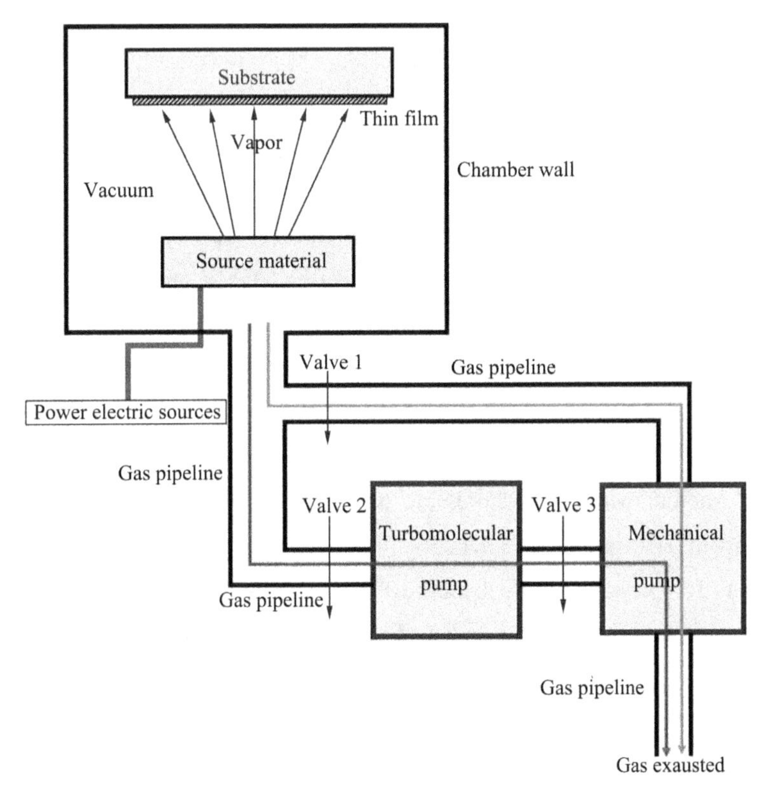

Figure 2.1 Schematic picture of a vacuum thin-film-deposition system, including the vacuum chamber and pumping systems

These pumps are generally used in series to make the vacuum systems work more effectively. This chapter describes the basic concepts of vacuums, vacuum acquisition, vacuum measurement, and the working principles of different pumping systems.

2.1 Basic concepts of vacuum

2.1.1 Definition and unit of vacuum

The word *vacuum* refers to a pressure well below that of a standard atmospheric gas (101,325 Pa). Here, molecular density is much lower than that at standard atmospheric pressure; thus, it is a rarefied gas state.

For a rarefied gas in equilibrium, the state of the gas can be described by the ideal-gas equation:

$$p=nkT, \tag{2-1}$$

or

$$pV=\frac{m}{M}RT, \tag{2-2}$$

where p in the formula is the pressure, n is the molecular density, V is the volume, T is the temperature, M is the molecular weight, m is the molecular mass, k is Boltzmann's constant, and R is the universal gas constant.

From Formula (2-2), we know that the molecular density of a gas at standard atmospheric pressure is about $n=3\times10^{19}$ cm^{-3}($p=1.01\times10^{5}$ Pa, $n=3\times10^{19}$ cm^{-3}). Thus, even in the high-vacuum state of 10^{-6} Pa, the density of gas molecules remains as high as 3×10^{8} cm^{-3}. Therefore, the so-called "vacuum" is relative, and an absolute vacuum does not exist.

The Pascal (Pa), the internationally standard pressure unit, is defined as 1 Pa = 1 Newton/m^2. Another widely used pressure unit is Torr (or mmHg). The relationship between the Pa and Torr is: 1 Pa$=7.54\times10^{-3}$ Torr. European countries also use millibars (mbar) and bars for pressure units, 1 bar$=10^{5}$ Pa.

The conversion coefficients of various vacuum units are shown in Table 2-1.

Table 2-1 The conversion coefficients (Coff) of various vacuum units

Unit	Coff			
	Pa	Torr	mbar	atm
Pa	1	7.5×10^{-3}	0.01	9.87×10^{-6}
Torr	133	1	1.33	1.32×10^{-3}
mbar	100	0.75	1	9.87×10^{-4}
atm	1.01×10^{5}	760	1,013	1

2.1.2 Characteristics and applications of vacuum

Compared with an atmospheric environment, the vacuum condition has special characteristics. In a vacuum, the gas molecular density is low. Thus, the collision frequency between the molecules is reduced. The average distance traveled between two collisions of gas molecules is called the *mean free path*. In this case, the mean free path of the gas molecules increases. The collision frequency between gas molecules and the vacuum-chamber wall was reduced. Hence, the adsorption rate of gas molecules on the surface of wall is low. At the same time, the chemically active gas gradients (such as oxygen and water) decline. Thus, the vacuum state is a relatively clean environment.

The vacuum is usually classified according to the magnitude of pressures:

- The low vacuum: $10^5 \sim 10^2$ Pa;
- The middle vacuum: $10^2 \sim 10^{-1}$ Pa;
- The high vacuum: $10^{-1} \sim 10^{-5}$ Pa;
- The ultra-high vacuum: $<10^{-5}$ Pa.

In the low-vacuum state, the density of gas molecules is still high. Thus, collisions between molecules are still frequent, the mean free path of the gas is short, and thermal movement of the gas remains dominant. This vacuum region is usually used to obtain a pressure gradient during thin film depositions.

In the middle-vacuum state, the number of molecules is about 10^{13} cm^{-3} ~ 10^{16} cm^{-3}. Compared with the atmospheric state, the number of molecules in the middle-vacuum state has been greatly reduced and the gas's flow state changes gradually from viscous to molecular flow. In an electric field, charged particles in the gas will be accelerated. They will collide, ionize neutral gas molecules, and produce gas-discharge phenomena called *plasma*. During this process, the kinetic properties of the gas are obvious, the convection phenomena disappear, and the chemical reactions between residual gases are greatly weakened. The deposition of thin films happens mostly during this process.

In the high-vacuum state, the density of molecules can be as small as 10^{10} cm^{-3}. Thus, the interaction between molecules is very weak, and the mean free path of gas molecules is larger than that of the container. Most gas molecules will collide with the surface of the vacuum-chamber wall. In this case, the thermal conductivity and internal friction of the gas are independent of pressure. The background pressures of most of the thin-film-deposition systems fall in this area.

In the ultrahigh-vacuum state, the density of molecules is below 10^{10} cm^{-3}. Collisions between molecules are rare compared to those between gas molecules and the vacuum-chamber walls. Pure solid thin film surfaces can then be obtained. The molecular-beam-epitaxy (MBE)-deposition system works in this vacuum region.

In all cases, the highest background vacuum (lowest pressure) is preferred for obtaining a clean thin-film-deposition environment and thus the best quality of the thin film.

2.2 Dynamics of gas

2.2.1 The velocity distribution of gas molecules

When gases are in a rarefied state, they can be described by the kinetic theory of ideal gases.

When gas molecules are in a steady state of motion, the velocities of the molecules change constantly. In the steady state, the velocity distribution obeys the Maxwell distribution:

$$f(v) = \frac{4}{\sqrt{\pi}} \cdot \left(\frac{m}{2kT}\right)^{\frac{3}{2}} \cdot v^2 \cdot e^{-\frac{mv^2}{2kT}}. \tag{2-3}$$

According to Formula (2-3), several rates of gas can be obtained, including the most probable rate:

$$v_{\mathrm{m}} = \sqrt{\frac{2kT}{m}}; \tag{2-4}$$

the average speed:

$$\bar{v} = \sqrt{\frac{8kT}{\pi m}}; \tag{2-5}$$

and the root-mean-square rate:

$$v_r = \sqrt{\frac{3kT}{m}}. \tag{2-6}$$

2.2.2 The mean free path

The mean free path of gas molecules λ can be defined as:

$$\lambda = \frac{1}{\sqrt{2}\pi\sigma^2 n}. \tag{2-7}$$

This quantity is related to the density n and diameter σ of molecules. According to Formula (2-1), the upper form can be rewritten as

$$\lambda = \frac{kT}{\sqrt{2}\pi\sigma^2 p}. \tag{2-8}$$

This formula shows that the mean free path of a gas molecule is proportional to the temperature, and inversely proportional to the pressure. When the type and temperature of the gas are constant,

$$\lambda \cdot p = \text{constant}. \tag{2-9}$$

That is, the mean free path is determined only by pressure p. For example, for 25 ℃ air, we have:

$$\lambda = \frac{0.667}{p}(\text{cm}),$$

where p has units of Pa.

2.2.3 The collision frequency of gas molecules per unit area

The number of molecules colliding with a wall surface per unit time and unit area can be represented by frequency ν:

$$\nu = \frac{1}{4}n\overline{v}. \tag{2-10}$$

Inserting Formula(2-1) and (2-5) into the above equation, we have

$$\nu = \frac{p}{\sqrt{2\pi mkT}}. \tag{2-11}$$

For example, for the air at 20 ℃, $\nu = 2.86\times10^{18}p(/\text{cm}^2\text{s})$, where p is pressure.

The collision frequency between gas molecules and the unit surface also represents the incident flux of gas molecules per unit area. As an application example, we calculate the time required for a clean solid substrate surface to be contaminated by an impurity gas in the environment under vacuum. Assuming that every incident gas molecule is adsorbed onto the solid surface, the surface area of the substrate is covered by a layer of gas molecules within a time

$$\tau = \frac{N}{v} = \sqrt{2\pi mkT} \cdot \frac{N}{p}, \tag{2-12}$$

where N is the number of molecules per unit area on the surface of the substrate. Calculations show that the time that it takes for the surface of a clean substrate to be covered by a single layer of molecular gas in air is 3.5×10^{-9} s. If the substrate is placed in a vacuum of 10^{-8} Pa, the contamination time will be extended to about 2.8 hours. This indicates that the vacuum environment is necessary to maintain a "clean" surface.

2.3 Gas flow and gas extraction

2.3.1 The flow of gases

When a pressure difference exists within a space, the gas will generate macroscopic directional flow. The gas-flow state depends upon the geometric size of the vessel and the gas pressure, temperature, and type of the gas. The Knudsen number can be used to differentiate the flow state, which is defined as

$$K_n = \frac{\lambda}{D}. \tag{2-13}$$

In this formula, D is the size of the container and λ is the mean free path of the gas. According to the size of K_n, the flow state of the gas can be roughly divided into three different states:

- ◆ The molecular flow, $K_n > 1$;
- ◆ The viscosity and molecular flow, $1 > K_n > 0.01$;
- ◆ The viscous flow, $K_n < 0.01$.

In the high-vacuum state (low pressure), the mean free path λ of the gas molecules is much larger than that of the container size D, and the gas-flow state is molecular flow. In the low-vacuum state (high pressure), λ is very short. This case is called viscous flow. The state between molecular flow and viscous flow is called middle-viscous molecular flow, or occasionally transitional flow.

2.3.2 Flow guide of a gas pipeline

In a vacuum system, the components are joined together by pipes, through which the gas flows. The ability to pass gas through a pipe is called flow conductance (C) and is defined as

$$C=\frac{Q}{p_1-p_2}, \tag{2-14}$$

where p_1 and p_2 are the pressures at both ends of the pipe, whose unit is Pa and Q is the unit-time gas flow through the pipeline, whose unit is Pa · L/s. The unit of flow conductance C is L/s, and L is the volume.

The magnitude of flow conductance depends upon the flow state of the gas and the shape of the pipe. The following is a common formula for calculating the conductance C of a pipeline under molecular-flow conditions:

(1) The flow conductance of a long, circular tube ($\frac{L}{d}>20$) is

$$C_L=3.81\times10^6\sqrt{\frac{T}{M}}\cdot\frac{d^3}{L}(\mathrm{L/s}), \tag{2-15}$$

where L is the pipe parameter and d is the inner diameter of the pipe.

(2) The flow conductance of a hole is

$$C_0=3.64A_0\sqrt{\frac{T}{M}}(\mathrm{L/s}), \tag{2-16}$$

where A_0 is the hole's area.

(3) A short pipe ($\frac{L}{d}<20$) can be regarded as a series of long tubes and holes. The flow conductance is

$$C=\frac{C_L\cdot C_0}{C_L+C_0}. \tag{2-17}$$

In the formula, C_L and C_0 are the flow conductance of the long tube and hole, respectively.

(4) Calculation of flow conductance of pipe series and parallel connections:

The flow conductance can be calculated according to the electric conductance in the circuit. "1, 2,…" represents a series or parallel connection:

series:
$$\frac{1}{C}=\frac{1}{C_1}+\frac{1}{C_2}+\cdots; \tag{2-18}$$

parallel:
$$C=C_1+C_2+\cdots. \tag{2-19}$$

More pipe-flow-conductance formulas can be found in relevant professional manuals.

2.3.3 Pumping speed

To obtain a vacuum, a pump is used to remove the air from the vacuum chamber.

The pumping rate is defined as the flow of gas through the inlet of the pump at a given pressure per unit time:

$$S=\frac{Q}{p}. \tag{2-20}$$

The pumping speed has units of L/s or m^3/s. Pumping speed is a function of pressure.

In an actual vacuum system, a pump with speed S_p is connected to the vacuum through a conduit with a flow conductance C. Its real pumping speed (or effective pumping speed), S_e, is

$$S_e=\frac{S_p \cdot C}{S_p+C}. \tag{2-21}$$

Thus, S_e is smaller than the theoretical pumping speed S_p or C. S_e is also limited by the smaller of S_p and C. For example: $C=S_p$, $S_e=\frac{1}{2}S_p$.

The effective pumping speed will be reduced due to the restriction of flow in the pipe, valve and other parts. Under normal circumstances, one should choose an appropriate pipe flow conductance and other components in the exhaust-system design of a vacuum system to ensure full pumping speed; typically, the velocity loss of the vacuum pump will not be greater than 40%~60%. For the low-vacuum system, the flow velocity of the pipeline should be reduced to about 10%. To this end, a high-vacuum pipeline can be selected using stainless steel pipes. The diameter should be equal to the inlet diameter of the vacuum pump and the length as short as possible. The low-vacuum pipeline can be made of a stainless-steel tube or a PVC rubber tube, and the inner diameter of the pipeline should equal that of the pump inlet.

2.3.4 Extraction equation

Pumping a vacuum vessel with a vacuum pump will reduce the pressure in the container. At the same time, the gas adsorbed on the inner wall of the container will also be released into the container. The gas outside of the container will leak inside through openings such as the weld seam of the container, causing the internal pressure to increase. The dynamic variation of the gas pressure in the container can be described by the pumping equation

$$V \cdot \frac{dp}{dt}=-p \cdot S_e+Q, \tag{2-22}$$

where S_e is the effective pumping speed and Q is the amount of gas entering the container per unit time. This includes the amounts of outgassing on the wall surface and of air leakage.

When $Q<pS_e$, the pressure in the container decreases with the pumping time. When $Q=pS_e$, $\frac{dp}{dt}=0$, $p=$constant, the pressure in the vessel no longer varies; it reaches the ultimate vacuum or the ultimate pressure.

In order to achieve the high-vacuum limit, Q should be reduced as much as possible to reduce the amount of leakage and outgassing. The limit is $Q=0$, when the equation of gas extraction becomes

$$V \cdot \frac{dp}{dt}=-p \cdot S_e, \tag{2-23}$$

which yields

$$\frac{dp}{p}=-\frac{S_e}{V} \cdot dt. \tag{2-24}$$

If the integral of the formula was taken at $t=0$ and $p=p_1$, the pressure at any time t will be:

$$p=p_1 \cdot e^{-(S_e/V) \cdot t}. \tag{2-25}$$

From this formula, when S_e/V is certain, p decreases as time goes on. In theory, $t \to \infty$, $p \to 0$. In fact, the effective pumping speed S_e is a function of pressure, which generally decreases along with p. When the pressure drops to a steady value (i.e. $\frac{dp}{dt}=0$), the vacuum system reaches the limit pressure at once. According to Formula (2-22), we have $p \cdot S_e=Q$. Therefore, the ultimate pressure is

$$p=\frac{Q}{S_e}. \tag{2-26}$$

Therefore, in order to obtain high or ultra-high vacuum, the effective pumping speed S_e should be increased as much as possible. At the same time, the air intake Q should be reduced.

 2.4 Vacuum acquisition

The vacuum environment requires the use of various pumps. According to the principle of pumping, vacuum pumps can be divided into two major categories, namely transport vacuum pump and capture vacuum pump. The former uses gas compression to transfer gas molecules to the vacuum system. The latter relies on the aggregation or adsorption of gas molecules in the vacuum system to capture the gas molecules and then excludes them from the system.

Transport vacuum pumps can be subdivided into mechanical gas-transfer pumps and air-flow pumps. Rotary mechanical vacuum pumps, root pumps and turbine molecular pumps are typical examples of mechanical gas-delivery pumps, while oil-diffusion pumps are air-driven gas pumps. Capture vacuum pumps are subdivided into low-temperature-adsorption pumps and sputtering-ion pumps. The working pressure ranges of varies pumps are listed below in Figure 2.2:

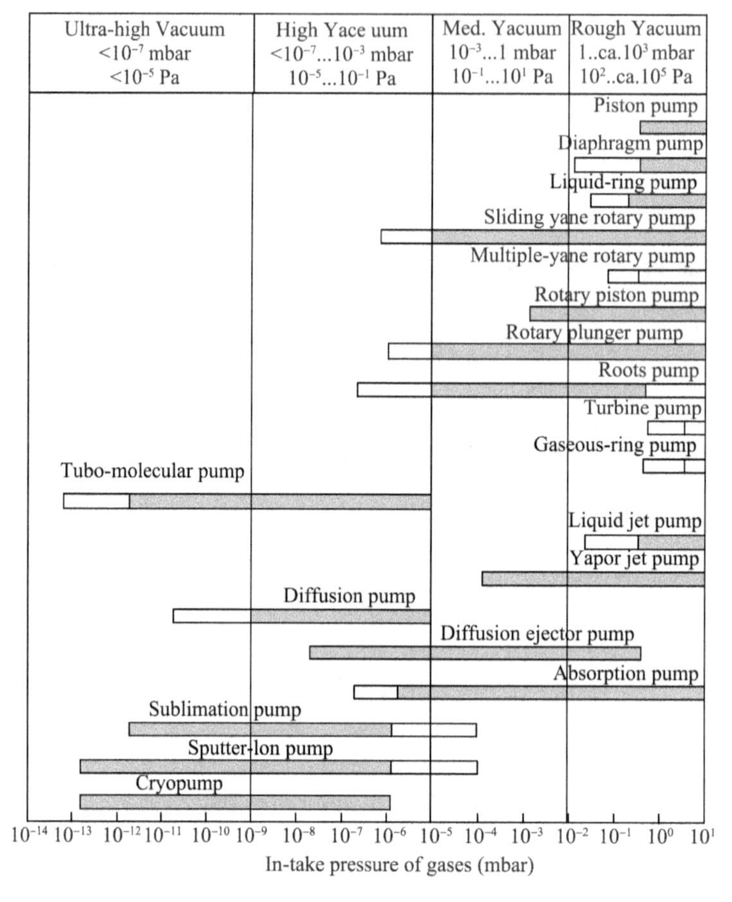

Figure 2.2 Schematic picture of working pressure ranges of varies pumps[19]

2.4.1 Rotary mechanical pump

Rotary-piston vacuum pumps (also known as reciprocating vacuum pumps) and rotary-vane vacuum pumps are the two most widely used rotary mechanical vacuum pumps. As shown in Figure 2.3, the spool valve is driven by an eccentric linkage. When the slide valve moves, the gas is sucked into the pump body through the suction port and then discharged through the exhaust valve under compression of the slide valve.

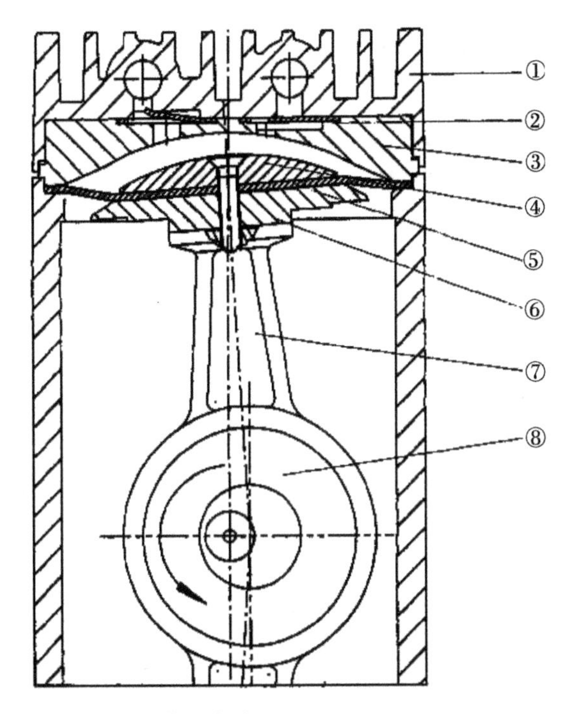

Schematic section drawing of a diaphragm pump stage
① Housing ② Valves ③ Head cover ④ Diaphragm clamping disc ⑤ Diaphragm
⑥ Diaphragm supporting disc ⑦ Connecting rod ⑧ Eccentric bushing

Figure 2.3 Schematic picture of the slide-valve pump[1,13]

Figure 2.4 shows a schematic of a rotary-vane vacuum pump. It includes four working stages: gas induction, gas isolation, gas compression and gas exhaust.

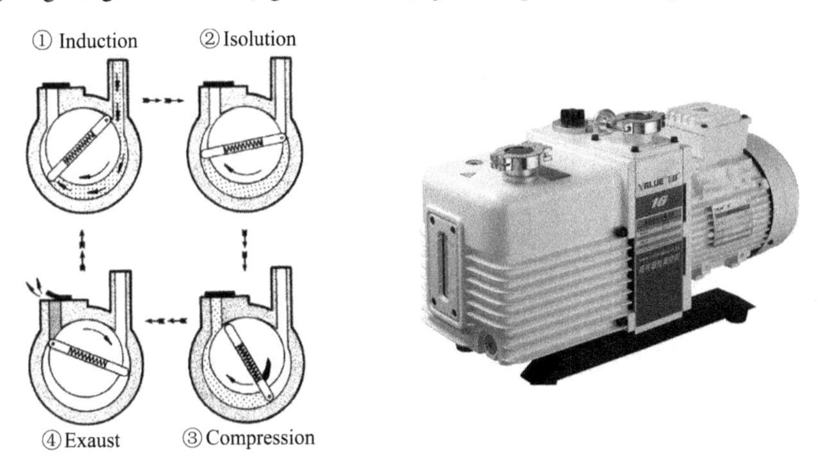

Figure 2.4 Schematic picture of the four working stages of rotary-vane pump (RVP) and an example picture of RVP[15]

(1) *Induction*: The gas molecules is introduced into the inlet of the pump;

(2) *Isolation*: The gas molecules is isolated by the crescent shaped volume from both the inlet and the outlet;

(3) *Compression*: The gas molecules is "pushed" to the outlet portion of the vacuum pump while outlet valve is still closed;

(4) *Exhaust*: As the rotor pushes the gas into the tiny outlet valve, pressure builds up; the high pressure forces the outlet valve to open and gas is discharged.

To improve the sealing effect of the gas and prevent backflow, there is only a small fit gap between the two pumps' moving parts, and oil is used as the sealing substance between moving parts. Oil also plays a role in the lubrication of these mechanical components.

The pumping speed of a rotating mechanical pump can be calculated as follows. In each pump-rotation period, the gas with volume V between the rotor and stator is completely excluded from the pump. At this time, the theoretical pumping speed should be equal to that of the pump,

$$S_p = Vf, \tag{2-27}$$

where f is the speed of the pump. The pumping speed of the spool-valve vacuum pump is generally 10~500 L/s, while that of the rotary-piston vacuum pump is about 1~300 L/s. The pumping speed of the two pumps is determined by Formula (2-27) when the pressure is relatively high; when the pressure decreases gradually, the pumping speed will gradually decrease. The limiting vacuum degree of a single-stage rotary piston pump can reach 10^3 Pa, and the two-stage spool valve can be down to 1 Pa after serial connection. *For the rotary-vane pump, the limit vacuum of single-stage operation reaches* 10^{-1} Pa, while the two such pumps working together can decrease the limit-vacuum pressure to 10^{-2} Pa.

The two mechanical-vacuum pumps can be used alone or as the first stage for other vacuum pumps. The pumping speed of a mechanical pump ranges from several L/s to tens of L/s. Generally, it can work in the pressure range from atmospheric pressure to several Pa.

2.4.2 Roots pump

Another form of the mechanical gas-delivery pump is the *Roots pump*, as shown in Figure 2.5. In the working condition, two rotors in the shape of an "8" rotate in opposite directions. The rotor has a high precision of articulation, so the oil is no longer used as the sealing medium between the rotor and the pump body. The space of rotor rotation during each sweep is very large, and the symmetry of the rotor pump is also high. Thus, the rotor can rotate at very high speed. The pump speed can be large (such as 10^3 L/s) and the vacuum limit can reach below 10^{-3} Pa. However, the speed of pump will decrease when the pressure is below 10^{-1} Pa. It will also decrease rapidly when the pressure is above 2,000 Pa. Therefore, this pump is generally used in series with rotary mechanical pumps.

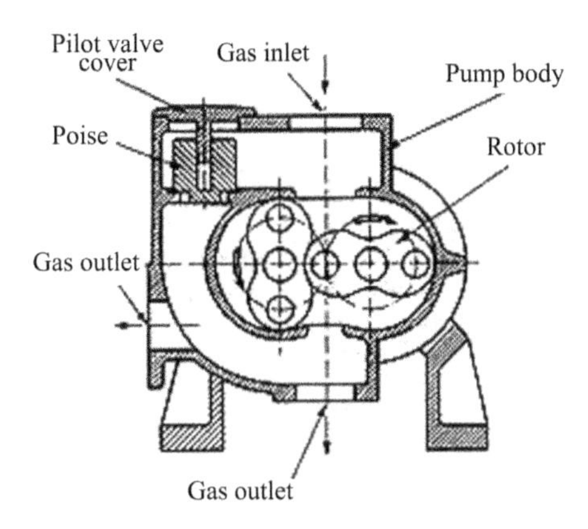

Figure 2.5 Schematic picture of a roots pump[13]

2.4.3 Oil-diffusion pump

A schematic of the oil-diffusion pump is shown in Figure 2.6. The working principle of this pump is to heat pump oil, which is stored at the bottom of the pump body, to a high-temperature evaporation state (about 200 ℃). When oil vapor is ejected from the multi-stage nozzle at high speed and directionality, the momentum is transferred to the gas molecules. Then, these gas molecules are forced to move towards the exhaust port, where they are compressed and then are discharged from the pump body. At the same time, the oil vapor, which is cooled by the pump body, condenses and returns to the bottom of the pump. The working principle of the oil-diffusion pump means that it can only be used in the vacuum state between 1 and 10^{-8} Pa, and cannot be directly connected with atmosphere. Therefore, before using an oil diffusion pump, various mechanical pumps are required for pre-vacuuming up to about 1 Pa. The pumping speed of the oil-diffusion pump ranges from several liters per second to 10^4 liters per second. One drawback of the oil-diffusion pump is that the return of oil vapor in the pump directly results in contamination of the vacuum system. Hence, oil-diffusion pumps are generally not used in precision analytical instruments or other ultra-high vacuum systems. In addition, when the oil pollution is not very high, a cold trap can be added between the oil diffusion pump and the vacuum chamber to let most of returned oil vapor be condensed instead of diffusing into the vacuum chamber. However, one problem this causes is that the flow conductivity of the system and the effective pumping speed of the pump will be reduced.

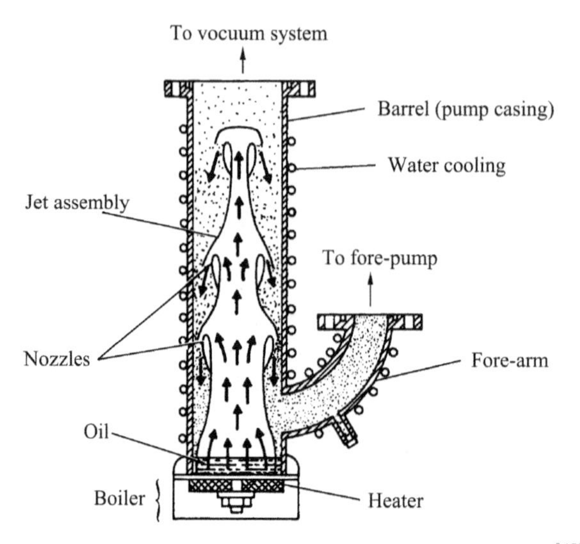

Figure 2.6 Schematic picture of an oil-diffusion pump[15]

2.4.4 Turbo-molecular pump

The turbo-molecular pump is an oil-free high vacuum pump, as shown in Figure 2.7. The rotor blades of a turbo-molecular pump have a particular shape, such that when they rotate at a speed of 20,000 ~ 30,000 r/min, the blades pass their momentum to the gas molecules. Because of the turbo-molecular pump with multi-stage blades, the gas molecules

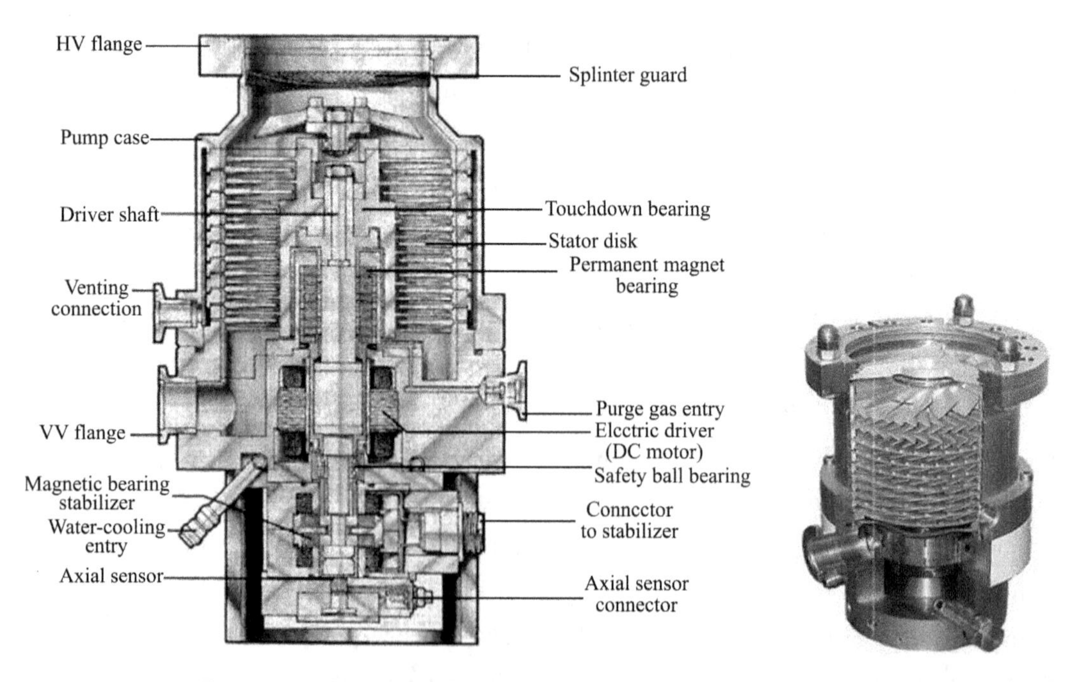

Figure 2.7 Schematic picture of a turbo-molecular pump (TMP) and an example picture of TMP

are transported from the upper-level blades to the next-level blades and continue to be compressed to the next level. Therefore, one characteristic of the turbo-molecular pump is that it is very effective for removing general gas molecules. For example, for nitrogen, the compression ratio (i.e. the ratio of the outlet pressure to the inlet pressure) can reach 10^9. However, the turbo-molecular pump's ability to extract gases with low atomic mass is poor. For example, hydrogen has a compression ratio of only about 10^3.

Due to the high compression ratio of the gas, the oil-vapor reflux problem can be ignored. The limit vacuum degree of the turbo-molecular pump can reach 10^{-8} Pa, the pumping speed can reach 1,000 L/s, and the pressure range is between 1 and 10^{-8} Pa. However, it cannot work in an atmospheric pressure environment. Thus, *a rotary-vane mechanical pump must be used as its front stage*, as shown in Figure 2.1.

2.4.5 Low-temperature adsorption pump

Low-temperature adsorption pumps extract gas molecules by spontaneous condensation or adsorption on the surfaces of other substances at low temperature, so as to obtain high vacuum. The vacuum degree obtained by this method depends on the temperature, the surface area of the adsorbed material, and the type of adsorbed gas. The vacuum degree is generally between 10^{-1} and 10^{-8} Pa. Figure 2.8 shows a schematic diagram of a cryogenic-adsorption pump driven by a circulating refrigerator. To reduce the heat exchange between the cryogenic chamber and the outside world, liquid nitrogen is also used as a thermal-isolation layer.

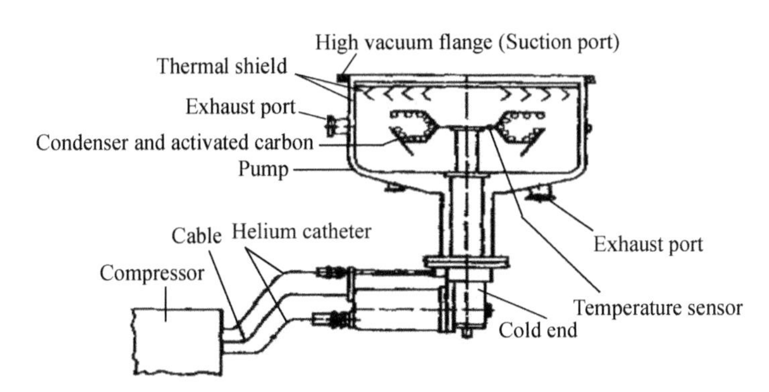

Figure 2.8 Schematic diagram of the low-temperature adsorption pump[1,2,13]

The materials used as adsorptive surfaces for gases often include: (1) metal; (2) a low-temperature material covered by a condensate of high-boiling-point gas molecules such as Ar and CO_2 (one such example is the adsorption of H_2 and He molecules on a low-temperature surface covered with Ar and CO_2 molecules); (3) adsorption materials with a large specific surface area, such as activated carbon, zeolite, etc.

The pre-vacuum required by the low-temperature adsorption pump should be less than 10^{-1} Pa, which can reduce the pump's heat load and avoid the accumulation of too-thick gas-condensation products on the pump's body. The vacuum limit of the low-temperature adsorption pump is related to the type of gas extracted. At equilibrium, the rate of gas-molecule acceptance on the condensation surface in the pump is equal to the evaporation rate of gas molecules on the surface of the vacuum chamber:

$$p_0 = p_s(T)\sqrt{\frac{300}{T}}. \tag{2-28}$$

We assume that the temperature on the surface of the vacuum chamber is 300 K. T is the temperature of the condensing surface in the pump and p_s is the vapor pressure of the extracted gas. For example, when the nitrogen is at 20 K, the vapor pressure is about 10^{-9} Pa, so the corresponding vacuum limit of the cryogenic pump is about 5×10^{-9} Pa.

According to Formula (2-28), gases such as H_2, He, and Ne, which have higher vapor pressures at low temperatures, are not easily removed by cryogenic-adsorption pumps. In addition to the above mentioned gases, the adsorption rate of the cryogenic pump is very large; this is because it depends only upon the velocity of gas molecules moving toward the condensation surface, as well as the surface area of the condensing process. Low-temperature adsorption-pump operating costs are relatively high; however, as a method of obtaining an oil-free high vacuum environment, it can be used in combination with other high-vacuum pumps such as turbo-molecular pumps.

2.4.6 Sputtering-ion pump

The sputtering-ion pump is shown in Figure 2.9. Its working principle is that the electron beam emitted from the Ti cathode ionizes the residual gas molecules and the ions produced will sputter out a large number of Ti atoms when they hit the cathode at high speed. These sputtered Ti atoms are deposited onto the surface of the inner wall. Due to the high activity of Ti atoms, they will capture gas molecules by physical or chemical adsorption and deposit them into the pump body, thus realizing a high-vacuum environment.

Obviously, the sputtering-ion pump speed depends on the gas. It is higher for gases with higher activity. For example, the pumping speed of a sputtering-ion pump for H_2 is several times that of O_2, H_2O vapor, or N_2. Due to the continuous sputtering of the Ti cathode, the life of an ion pump is limited. To prolong its service life, the sputtering pressure of the sputtering ion pump is set to 10^{-4} Pa and its vacuum limit can reach about 10^{-8} Pa.

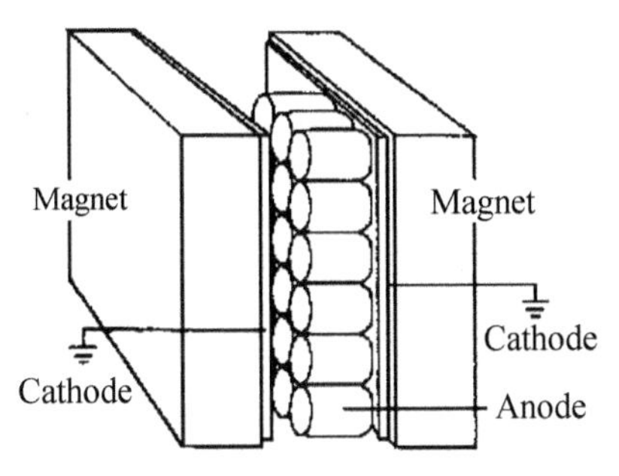

Figure 2.9 Schematic picture of the sputtering-ion pump[13]

2.5 Vacuum measurement

The measurement of vacuum degree is mainly a measurement of pressure. The measurement method differs according to the range of vacuum degree or gas pressure. Two common vacuum-measurement methods are described below.

2.5.1 Thermocouple vacuum gauge and Pirani vacuum gauge

The sensor used in vacuum measurement is often called a vacuum gauge. Both thermocouple vacuum gauges and Pirani vacuum gauges have been designed based on the change of the gas's thermal conductivity with its pressure. They are the most commonly used pressure-measurement tools in low vacuum.

As shown in Figure 2.10(a), in the thermocouple-vacuum gauge, a hot Pt wire is suspended and a constant current is passed through it. After the heat balance is achieved, the heating power provided by the current is equal to that lost through thermal radiation, metal-wire heat conduction, and gas-molecular heat conduction. Therefore, the equilibrium temperature of the hot wire will change regularly with vacuum degree, as shown in Figure 2.10(b).

For pressures of 0.1 ~ 100 Pa, the thermal conductivity of a gas will increase with its pressure. Hence, the temperature of the hot wire will decrease. At this time, the thermocouple is used to measure the temperature of the hot wire itself, and the ambient gas pressure is measured accordingly. Above this pressure range, the relationship between the couple's electromotive force and pressure becomes uncertain. Therefore, the thermocouple vacuum gauge cannot be used to measure vacuum degrees that are too low or too high. When the gas pressure exceeds 100 Pa, the thermal conductivity of the gas will not change with the gas pressure. At this point, the sensitivity of the method of measuring gas pressure with hot-wire temperature will rapidly decrease. When the gas pressure is below 0.1 Pa, if the proportion of heat carried away by gas-molecular conduction in total heating power is too small, then measurement sensitivity will also decrease.

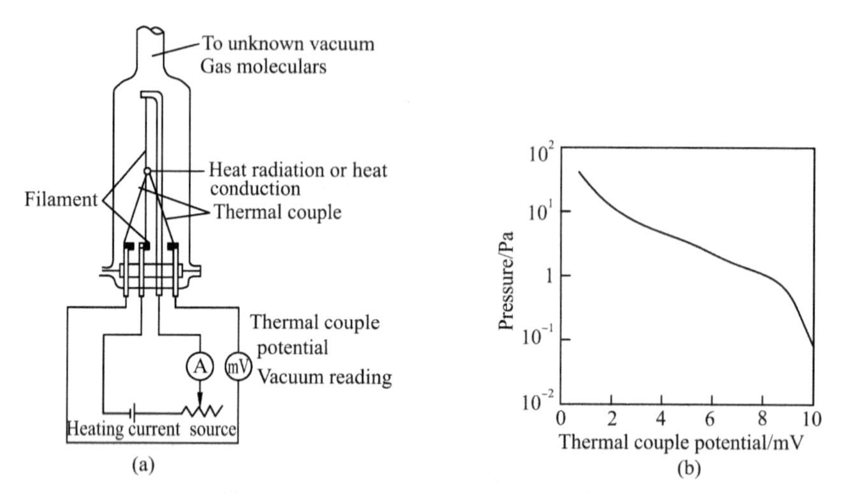

Figure 2.10 Schematic picture of a thermocouple vacuum gauge and variation law of the thermoelectric potential of thermocouple vacuum gauge with gas pressure

This method also has some additional shortcomings, such as a non-linear indication value in the measurement range, measurement results related to gas type, serious zero drift, and so on. The merits of this method are that it has a simple structure and is convenient for use.

A Pirani vacuum gauge is also called a thermal-resistance vacuum gauge. Its working principle is similar to that of the thermocouple vacuum gauge, giving it similar advantages and disadvantages. The Pirani gauge can measure vacuum degree as the hot-wire resistance with temperature. For different measuring-circuit settings, the Pirani vacuum gauge can be divided into constant-temperature types, constant-current types, and constant-voltage types. The vacuum degree of the Pirani gauge can range from 0.1 Pa to 0.1 MPa.

2.5.2 Ionization vacuum gauge

Ionization vacuum gauge is most commonly used for high vacuum. Its structure is shown in Figure 2.11. The ionization vacuum gauge mainly consists of a cathode, an anode, and an ion collector. The electrons emitted from the hot cathode collide with the gas molecules and ionize them as they fly to the anode. By using the ion-collecting electrode to receive ions, the vacuum degree of the environment can be measured according to the ion current. The ion-current magnitude depends upon the electron-current intensity emitted by the cathode, the collision cross section of gas molecules, and the density of gas molecules. Therefore, in the case of fixed cathode-emission current and gas type, the ionic-current intensity will directly depend upon the gas pressure.

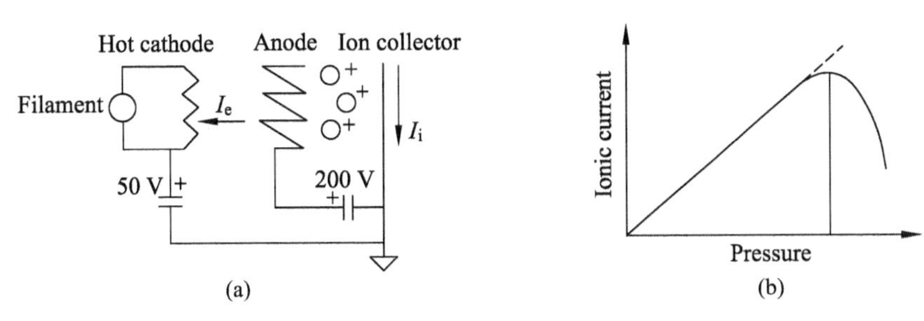

Figure 2.11 Schematic picture of the ionization vacuum gauge and the relationship between the ion current and the gas pressure of the ionization gauge

The lower limit of the ionization vacuum gauge's measurement range is called the X-ray limit value. It depends upon the photoelectric effect of the high-energy photons emitted from the cathode on the collector, as shown in Figure 2.11 (a). This effect produces a photocurrent equivalent to the ionic current at 10^{-9} Pa under vacuum conditions. The upper limit of the ionization vacuum gauge is about 1 Pa. At this time, the free path of electrons is too short to effectively ionize gas molecules. Between 10^{-9} and 1 Pa, the measured values of the ionization gauge are approximately linearly related to the gas pressure, as shown by the curve in Figure 2.11(b).

To eliminate the influence of the outgassing by the vacuum gauge upon the measurement at high vacuum, it is necessary to heat the gauge to a slightly higher temperature before using it, so as to reduce the outgassing of the electrode itself. Because the collision-ionization cross sections of different gas molecules vary, measurements by the ionization vacuum gauge also depend on the type of gas being measured.

Exercises

1. What is vacuum? Why is vacuum so important for thin film technology?
2. What are the basic units for pressure? How are they converted into each other?
3. How is the degree of vacuum classified?
4. What are the working-pressure ranges for a rotary-vane pump, a diffusion pump, and a turbo-molecular pump?
5. What is the working mechanism of the rotary-vane pump?

6. What is the working mechanism of the diffusion pump?
7. What is the working mechanism of the turbo-molecular pump?
8. What is the working mechanism of low-temperature-adsorption pump?
9. What are the working mechanisms and pressure-testing ranges of the thermocouple and Pirani vacuum gauges?
10. What are the working mechanism and pressure-testing range of the ionization vacuum gauge?

Chapter 3 Plasma Technology Basis

In our daily life, matter usually exists in solid, liquid or gaseous forms. In fact, there is the fourth state of matter—plasma, in which most of the matter in the universe exists. For example, lightning and auroras are plasma; the ionized gas in fluorescent and neon lights is plasma; the ionosphere above the earth is a low-temperature plasma; the sun and stars are high-temperature plasma; the interstellar medium, too, is a plasma.

In this book, the methods for preparing various thin film materials, including sputtering, plasma chemical-vapor deposition, and pulsed-laser chemical-vapor deposition, are all realized in the plasma. Basic knowledge of plasma is a foundation for thin film material preparation technology.

3.1 Basic concepts of plasma

Plasma can be formed by gas discharge. For example, if two metal electrodes are encapsulated in a glass tube, then a vacuum pump is used to reduce the air pressure to 13 ~ 133 Pa and a DC voltage is applied between the two electrodes. If the voltage rises slowly to hundreds of volts, the gas in the tube will become conductive. When the electric current increases sharply, and the discharge tube will emit orange-red light. Some neutral gas molecules in the gas will be ionized into ions and electrons, which are conductive. This ionized gas is called plasma. More strictly, plasma is a quasi-neutral macroscopic system composed of a large number of charged and neutral particles. Its properties arise from the collective behavior of these particles. Next, we will describe some basic plasma parameters.

3.1.1 Plasma temperature

Plasma contains electrons, ions, and a large number of neutral particles or groups. When the plasma system is in thermal equilibrium, it obeys the Maxwell distribution and its mean motion energy E is related to the root-mean-square (RMS) velocity as follows:

$$E=\frac{1}{2}mv^2=\frac{3}{2}kT. \tag{3-1}$$

Here, m is the particle mass, k is the Boltzmann constant, and T is the temperature. However, the low-pressure discharge plasma is often in a non-equilibrium state, and the temperatures of the electrons, ions, and neutral particles differ. Therefore, the plasma temperature should be classified as an electron temperature T_e, ion temperature T_i, and neutral-particle temperature T_n.

The plasma-energy state should be expressed as several temperatures, meaning that there is no thermal equilibrium between particles in plasma. This is because collisions between particles of the same kind are more frequent (i.e. it is easier to achieve thermal equilibrium between particles of the same kind). For example, the collision frequency between electrons is much higher than that between electrons and ions. In low-temperature plasma, the electron temperature is thousands of times higher than the ion temperature. This kind of plasma is called non-equilibrium plasma.

3.1.2 Plasma density

Plasma is composed of electrons, ions and neutral particles. In addition to gas plasmas with high electron affinity (such as oxygen and halogen gas), ions are usually positively charged. If the ion density n_i and the electron density n_e are used, then the following relationship is obeyed by the plasma:

$$n_e=n_i=n. \tag{3-2}$$

We call this the quasi-neutral condition, with n being the plasma density. Therefore, although plasma is composed of charged particles, it is electrically neutral as a whole and so is called plasma.

3.1.3 Plasma oscillation

The density distribution of particles in the plasma will fluctuate. If the electron in the plasma is displaced relative to the ion, the electron density in one place will increase while the ion density will increase in another place and an electric field will form in the plasma space. At this point, the electron will move back under the Coulomb force of the electric field; but because of inertia, the electron will cross the equilibrium position and be affected by the opposite Coulomb force again. Thus, the electron will oscillate around the equilibrium position at a certain characteristic frequency, which is called plasma oscillation. The angular oscillation frequency ω_p and frequency f_p can be expressed as follows:

$$\omega_p = \left(\frac{ne^2}{m\varepsilon_0}\right)^{\frac{1}{2}}, \tag{3-3}$$

$$f_p = \frac{\omega_p}{2\pi} = \frac{1}{2\pi}\left(\frac{ne^2}{m\varepsilon_0}\right)^{\frac{1}{2}}. \tag{3-4}$$

Plasma oscillation can be divided into plasma-electron oscillation and plasma-ion oscillation. The plasma-electron oscillation frequency is

$$\omega_{pe} = \left(\frac{4\pi n_e e^2}{m_e}\right)^{\frac{1}{2}}. \tag{3-5}$$

The oscillating frequency of ion oscillation is

$$\omega_{pi} = \left(\frac{4\pi n_i e^2}{m_i}\right)^{\frac{1}{2}}. \tag{3-6}$$

Because the electron mass m_e is very small, ω_{pe} is very high. For example, in the plasmas of $n_e = 10^{10}\ \text{cm}^{-3}$, $f_{pe} = 898\ \text{MHz}$, ω_{pe} belongs to the microwave-frequency band. The oscillation frequency of ion plasma is generally much lower than that of electron plasma.

3.1.4 Debye length

We know that when the electric charge is close to a metal conductor, electrostatic induction will occur. At this point, the electric charge moves towards the surface of the conductor so that the electric field within the conductor is zero. Plasma is conductive similar to that of an electric conductor. At a certain spatial scale, it can shield the electric field, or the Coulomb force of the charged particles in the plasma will have a limited range. This limit can be expressed by the Debye length λ_D,

$$\lambda_D = \left(\frac{\varepsilon k T_e}{n_e e^2}\right)^{\frac{1}{2}},$$

where λ_D decreases with the increase of electron density. When the geometrical linearity L of the system is $\lambda_D (L \gg \lambda_D)$, the ionized gas contained in the system can be regarded as plasma. At this time, the net charge is only within the Debye length, and the plasma outside is macroscopically neutral. The Debye length represents the spatial characteristic scale for maintaining the macroscopically neutral charge of the plasma.

3.2 Classification of plasma

Plasma can be classified according to their temperature and density, as shown in Figure 3.1.

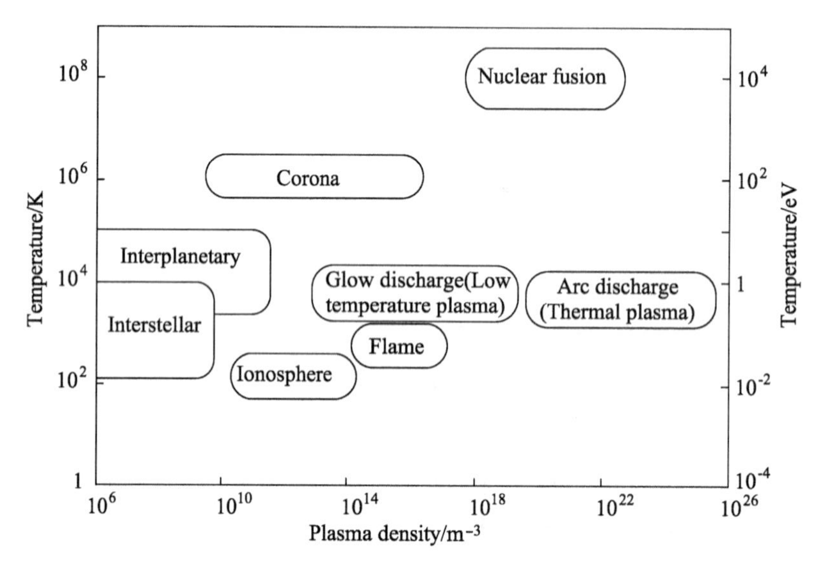

Figure 3.1 Classifications of plasma according to their temperature and density[13]

The electron temperature and density of natural plasma covers a wide range of parameters, as shown in the figure above. The density is between 10^6 and 10^{26} m^{-3} and the temperature is between $10^{-2} \sim 10^5$ eV. The interstellar medium consists of hydrogen plasma with a density of 10^6 m^{-3}. The ionosphere about 50 km away from the earth's surface is composed of plasma with a density of $10^{10} \sim 10^{14}$ m^{-3} and a temperature of 0.01 eV. The corona is a plasma with a temperature of several hundred eV, and the temperature of fusion plasma can reach more than 10 keV. Low-temperature plasma produced by glow discharge at low pressure is mainly used in the preparation of thin films.

3.2.1 Low-temperature plasma

In plasma chemical-vapor deposition and sputtering deposition, the low-pressure glow-discharge plasma is mainly used. The discharge pressure ranges from hundreds of Pa to 10^{-1} Pa. The discharge gives off a bright glow. The average voltage is high (from hundreds of V to thousands of V) and the current is small (from several mA to hundreds of mA). The plasma density is about $10^{10} \sim 10^{12}$ cm^{-3}. The electron temperature is about several eV. This is classified as a low-temperature plasma.

In general, the electron temperature of a cryogenic plasma does not reach thermal equilibrium with ions and gases; thus, they are also known as non-equilibrium plasmas. Figure 3.2 shows the temperature of the nitrogen plasma obtained by high-frequency discharge. When the pressure is several hundred Pa, the electron temperature T_e is above 12,000 K while the gas temperature T_g is only about 1,000 K. The general glow discharge in low-temperature plasma has the following relations with various temperatures:

$$T_e > T_i \sim T_g, \tag{3-7}$$

where T_i is ion temperature.

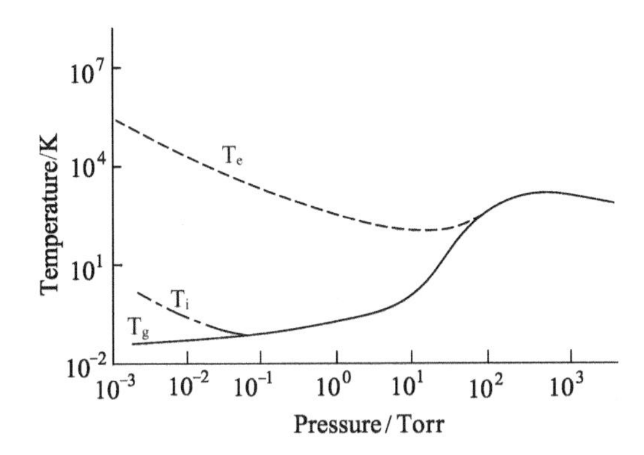

Figure 3.2 Relationship between pressure and temperature [13]

3.2.2 Thermal plasma

An arc discharge is generated at high pressure, characterized by low voltage (several V to tens of V) and high currents (several A to hundreds of A). In arc-discharge plasmas, the collisions of electrons with ions and gas particles are high, and the temperatures of various particles tend to balance, i.e. $T_e \sim T_i \sim T_g$. The changes of T_e, T_g, and T_i with pressure and electron density are shown in Figure 3.3 and Figure 3.4. When the pressure is higher than 10^4 Pa, the electron density can exceed 10^{15} cm^{-3}. This neutral-gas particle is also heated by a plasma called a thermal plasma.

In an arc discharge occurring at atmospheric pressure, if the plasma is cooled by an outer wall or gas, the surface of the plasma will decrease due to heat loss. In addition, the center of the plasma is concentrated, producing a thermal-contraction effect. High temperatures above 10,000 K can be obtained at the center of the plasma. This is a typical thermal plasma, which can be used for the sintering of ceramics and high-melting-point compounds, growth of single crystals, thermal spraying, welding, melting and other applications.

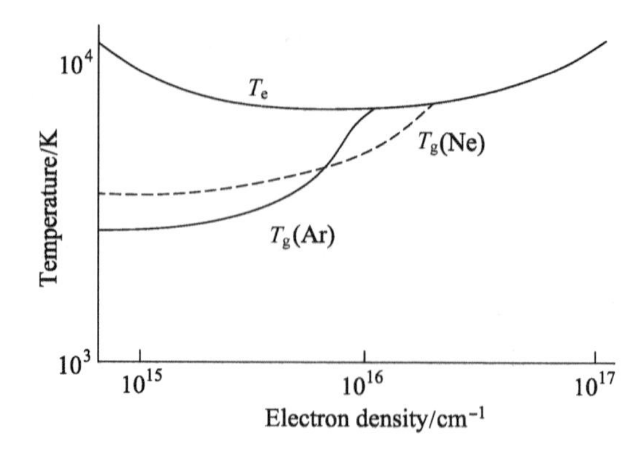

Figure 3.3 Relationship between electron density and temperature [13]

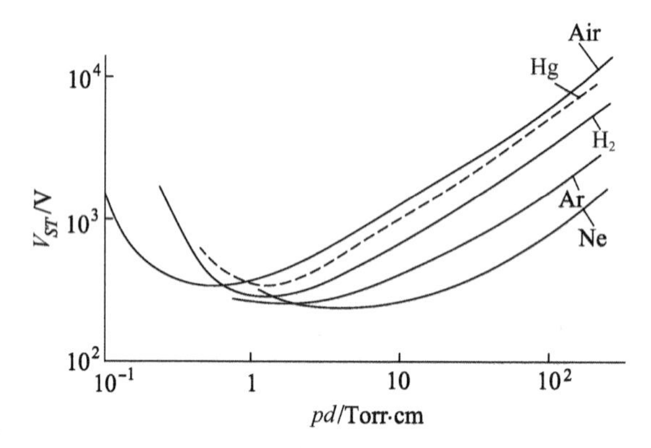

Figure 3.4 Relationship between the discharge-breakdown voltage *V* and *pd* of several gases[13]

3.3 Occurrence of low-temperature plasma

Low-temperature plasma can be produced at low pressure using DC, AC, RF, microwave, and light-wave electric fields to stimulate gas discharge.

3.3.1 Discharge breakdown

3.3.1.1 DC-discharge breakdown

A voltage V is applied between two parallel-plate electrodes at d, assuming that the inter-electrode field is uniform and the value is V/d. According to the theory of Townsend gas discharge, the function of the initial or breakdown voltage V_B, the product of the pressure p and the electrode spacing, and the functional relationship of pd are shown below:

$$V_B = \frac{Bpd}{\ln(pd) + \ln \dfrac{A}{\ln\left(1+\dfrac{1}{\gamma}\right)}}. \tag{3-8}$$

Here, $A = \frac{1}{\lambda_{el}}$, $B = \frac{u_t}{\lambda_{el}}$, λ_{el} is the average free path of electrons at a pressure of 1 Torr, u_t is the ionization potential of a gas, α is the gas-ionization coefficient, and γ is the two-electron emission coefficient of the ion-bombarding cathode. The relationship between the V and pd values of several gases is shown in Figure 3.4 and called the Paschen curve.

The breakdown voltage is related to the pd product. When the pressure is too low, the average free path of electrons, λ_e, is large. Most electrons and gas molecules do not collide until they reach the anode; thus, the breakdown voltage is higher. When the pressure is too high, the average free range of electrons is small. Electrons and gas molecules collide frequently, losing energy, and the electrons do not have enough energy to ionize the gas molecules. Therefore, the higher the pressure, the higher the breakdown voltage V_B.

3.3.1.2 High-frequency discharge breakdown

Under a high-frequency electric field $E\sin(\omega t+\theta)$, a particle with a mass of m and a charge e produces a round-trip motion between the electrodes, and its amplitude A is given as follows:

$$A = \frac{E}{\omega\sqrt{\left(\dfrac{1}{u}\right)^2 + \left(\dfrac{m}{e}\right)\omega^2}}. \tag{3-9}$$

Here, u is the mobility of charged particles. Formula (3-9) can be written as follows if the effects of particle inertia are neglected:

$$A = \frac{uE}{\omega}. \tag{3-10}$$

If the distance between the electrodes is d when $2A$ is greater than $d(2A>d)$, the charged particles will reach the electrode; when $2A$ is less than $d(2A<d)$, the charged particles will reciprocate in the space between the electrodes. This is called trapping particles.

If the result is expressed in $\frac{2A}{d} = \frac{2uE}{\omega d}$, then the breakdown voltage at high frequency is a function of pd, ωd, and fd:

$$V_B = f(pd, fd). \tag{3-11}$$

When pd is constant, the breakdown voltage is a function of fd. For example, when

$pd = 1$ Torr · cm, the dependence of the breakdown voltage upon the frequency is shown in Figure 3.5.

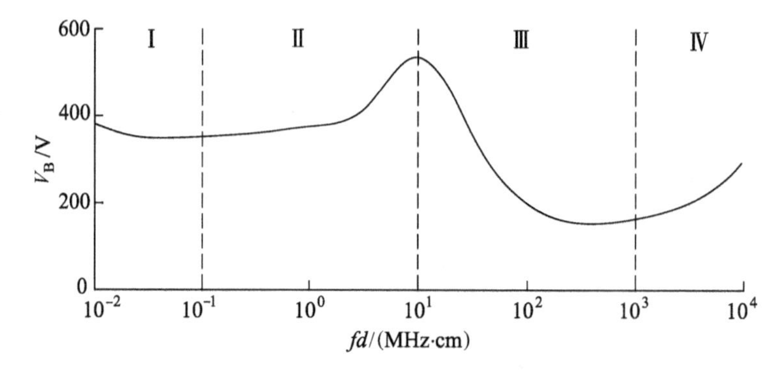

Figure 3.5 Dependence of the discharge breakdown voltage on frequency of air

Zone I is of low frequency and the characteristics are similar to those of DC discharge. Zone II is of intermediate frequency: here, the larger mass of ions cannot keep up with the electric field and they therefore become stuck in the discharge space. This leads to a decrease in the number of ions in the bombarding cathode and an increase in the breakdown voltage. Zone III is of high frequency: electrons are trapped in the discharge space due to an increase in the electric-field frequency, and collisions with gas molecules frequently ionize them, resulting in a sharp drop in breakdown voltage. Zone IV is of very high frequency; the potential of the electric field is backward due to the inertia of the electron and the bit of its migration velocity. The electrons do not get enough energy from the electric field to ionize the gas molecules; thus, the breakdown voltage rises gradually.

Electrons produced in the discharge space, on the one hand, move back and forth between electrodes in the presence of high-frequency electric fields; on the other hand, diffusion will cause losses on the electrode or tube wall. Therefore, the condition of sustained discharge in the high-frequency region is one of balance between electron generation and disappearance:

$$\frac{d}{\pi}\left(1-\frac{2B}{\omega d}\right)\sqrt{\frac{A}{D}}=1. \tag{3-12}$$

Here, A represents the ionization coefficient, B represents the electron's electric-field amplitude, and D is the electron diffusion coefficient.

3.3.2 DC glow discharge

3.3.2.1 Plasma sheath

In a discharge plasma, electrons move much faster than ions because they are much lighter in mass. Hence, electrons will first arrive at the wall after the discharge begins, and a negative-charge wall will be formed when a negative wall potential occurs. This potential will block electrons and accelerate ions. When the electron current and the ion current are equal, the total current will be zero and the equilibrium state will have been reached. Here, the potential of the wall is called a floating potential or a wall potential, expressed by V_f. In general, this is negative compared with the plasma potential V_a. The potential distribution of V_f is shown in Figure 3.6.

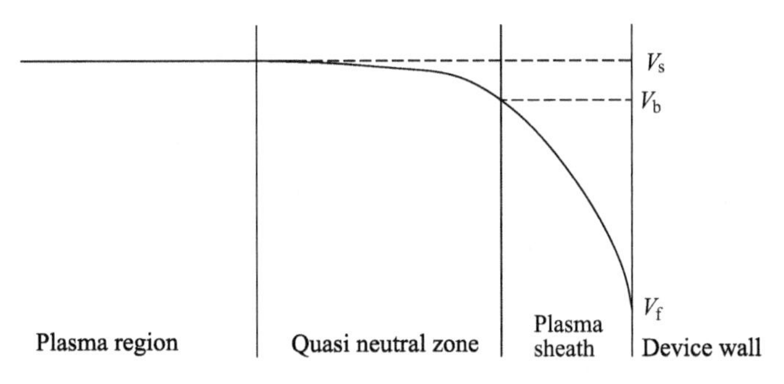

Figure 3.6 Potential distribution in the sheath[13]

The space potential interval between the device wall and the plasma region is called the plasma sheath. In fact, there is a transition zone or a neutral region between the sheath and the plasma region. A weak-current field exists in this region. After the ion is accelerated, it enters the sheath at a certain initial velocity. If the transition-zone potential is V_b, we have

$$V_a - V_b = \frac{kT_e}{2e}. \tag{3-13}$$

The electron and ion densities are n_e and n_i, respectively. The electron number in the unit wall area per unit time is $\frac{n_e v_e}{4}$, and the ion number is $\frac{n_i v_i}{4}$. Because $n_e = n_i$ in the boundary between the transition region and the sheath layer, we have

$$V_b - V_f = \frac{kT_e}{e}\ln\left(\frac{n_e v_e}{n_i v_i}\right) \tag{3-14}$$

$$= \frac{kT_e}{e}\ln\left(\frac{MT_e}{mT_i}\right). \tag{3-15}$$

Here, M and m are the masses of ions and electrons, respectively, and T_e and T_i are their temperatures. The average electron speed is $v_e = \left(\frac{8kT_e}{\pi m}\right)^{\frac{1}{2}}$. When the ion velocity in the sheath is approximately $v_i = \left(\frac{kT_e}{M}\right)^{\frac{1}{2}}$, Formula (3-15) can also be written as follows:

$$V_b - V_f = \frac{kT_e}{2e} \ln\left(\frac{8M}{\pi n}\right). \tag{3-16}$$

3.3.2.2 Normal and abnormal glow discharge

The current voltage characteristics of the glow discharge are shown in Figure 3.7. In the normal glow-discharge region, the current can vary by several orders of magnitude. However, the cathode drop indicated by the ordinate is little changed. The normal glow-discharge current increases with the cathode area from 10^{-3} A to 10^{-2} A. The cathode is then covered with plasma. To obtain a larger current, the cathode potential drop should be increased, such that the number of electrons emitted from the cathode increases and electrons goes into the abnormal glow discharge region. In the area with increasing discharge voltage, current rapidly increases. Therefore, a very high power density can be obtained at the cathode.

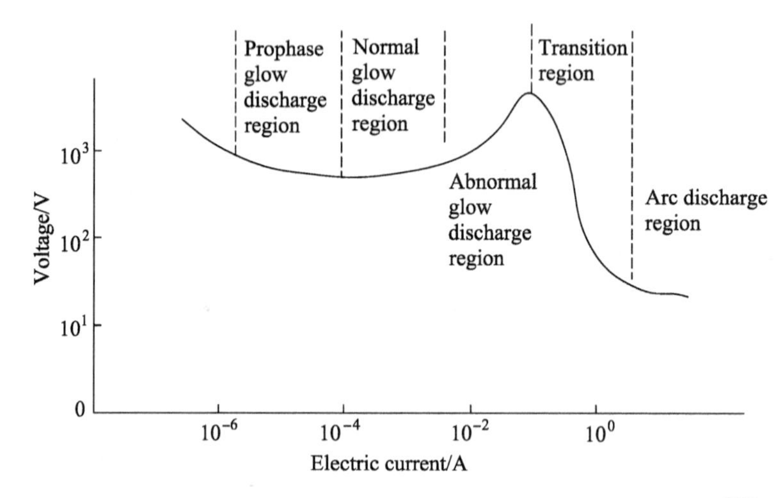

Figure 3.7 Current and voltage characteristics of glow discharge[13]

3.3.3 High-frequency discharge

A glow plasma can also be produced using a high-frequency electric-field discharge, as shown in Figure 3.8. If the power-supply load and discharges from the inside, it is called the inner-electrode type, as shown in Figure 3.9. If made of a quartz glass tube, the power-supply load and discharge from the outside is called the external-electrode type.

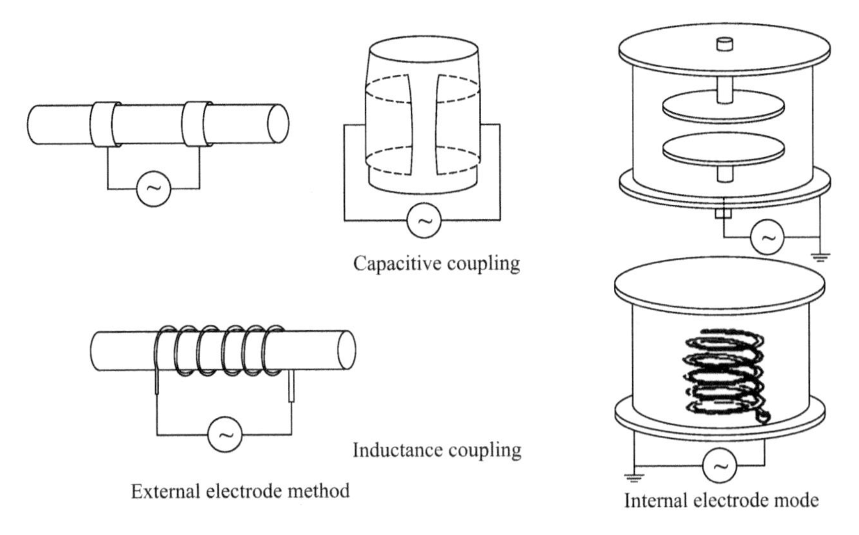

Figure 3.8 Generation of high-frequency discharge[13]

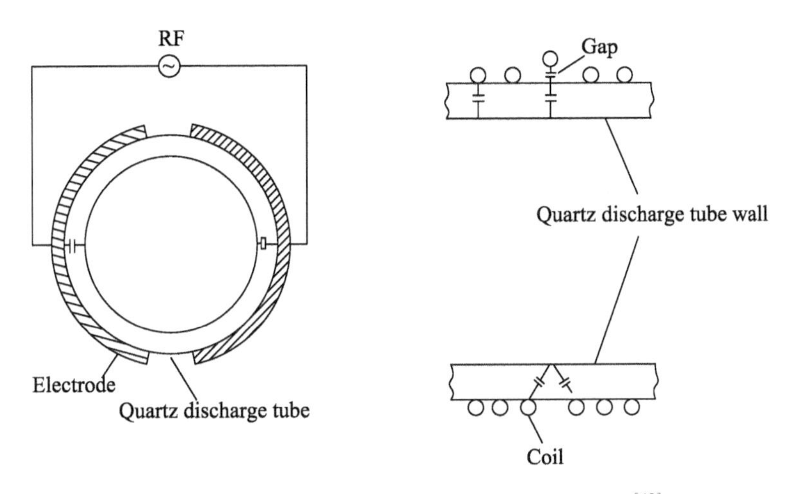

Figure 3.9 Capacitor and inductance coupling[13]

A high-frequency plasma system has two power-coupling modes: inductive or capacitive, as shown in Figure 3. 10. The commonly used discharge power-supply frequency is 13.56 MHz, which is in the radio-frequency area; therefore, this is often called radio-frequency discharge. To obtain impedance matching, the high-frequency power supply is connected to the load via a matching circuit in Figure 3.10. For a 50 ~ 400 kHz low frequency, we mainly use a step-up transformer. For a 13.56 MHz high frequency, a π shape-matching circuit is used.

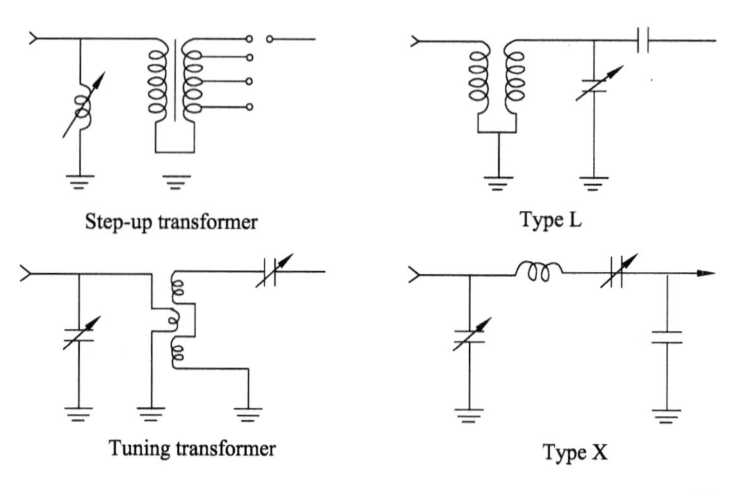

Figure 3.10 Matching circuit in a high-frequency plasma system[13]

Exercises

1. Please explain the concept of plasma temperature, plasma density and plasma oscillation.
2. What's plasma Debye length?
3. How is plasma classified?
4. Please explain the relationship between plasma techniques and thin film deposition techniques.

Chapter 4 Physical Vapor Deposition of Thin Films

Physical vapor deposition is a physically based method for preparing thin film materials. It includes many methods like thermal evaporation, ion-beam evaporation, sputtering, and pulsed-laser deposition, as shown in Figure 4.1. This chapter describes the basic principles and techniques for preparing films using these methods.

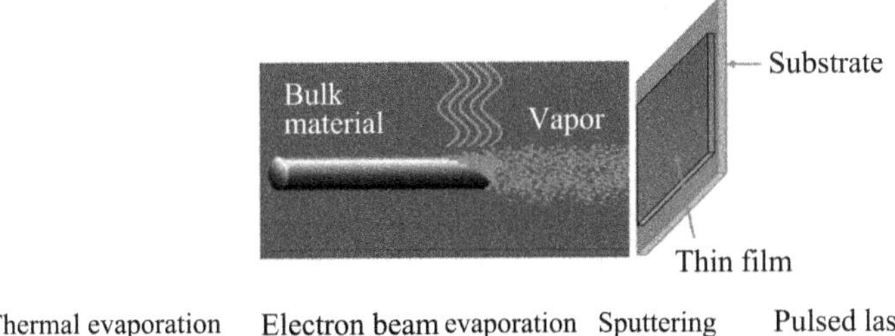

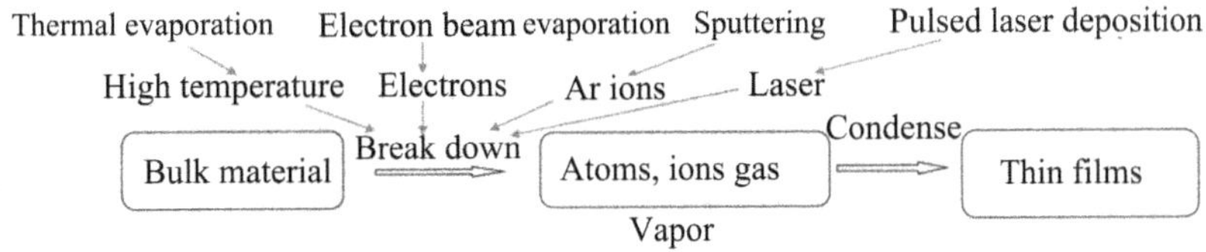

Figure 4.1 Schematic picture of physical vapor-deposition methods, including thermal evaporation, Electron beam evaporation, sputtering, and pulsed-laser deposition

4.1 Evaporation deposition

Evaporation is a common physical phenomenon and evaporation deposition is a common coating technology. The basic process of film evaporation deposition is:

(1) The raw materials are heated, evaporated and vaporized;

(2) The vaporized atoms or molecules are transported from the evaporation source to the substrate surface;

(3) The evaporated atoms or molecules are adsorbed, nucleated, grown on the substrate surface, and finally formed into a continuous film.

This section will introduce the basic principle and technology behind three aspects of evaporation deposition: evaporation of raw material, transport of vapor-phase particles and growth of the thin film.

4.1.1 Thermal-resistance evaporation source

The components that heat raw materials and gasify them are called evaporation sources. The most common heating methods are the resistance, electron-beam, and high-frequency-induction methods.

Thermal-resistance evaporation starts with a high-melting-point metal or ceramic material, which is then made into an evaporator of proper shape and loaded with raw material. The material is heated directly or indirectly using the electric resistance of the evaporator. The evaporation source has a simple structure and is a common source of evaporation. A schematic picture of the thermal-evaporation-deposition system with a resistance evaporation source is shown in Figure 4.2.

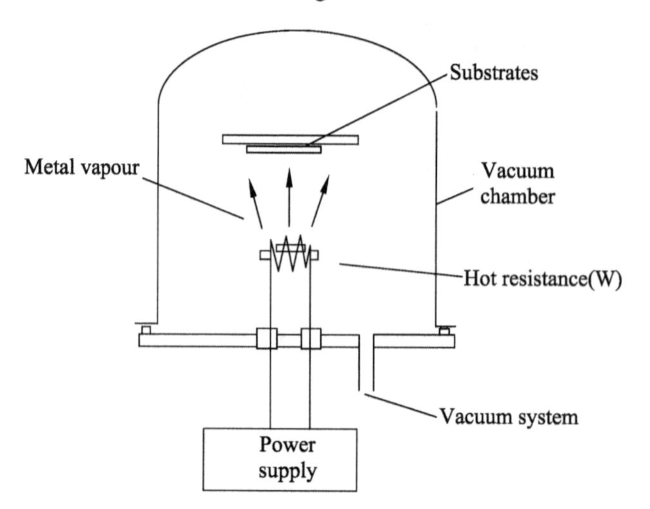

Figure 4.2 Schematic picture of the thermal-evaporation-deposition system with a resistance evaporation source

The main factors to be considered when making the evaporation source are its material and shape.

(1) *High melting point.* The evaporation temperature of most raw materials (the temperature of the saturated vapor is about 10^{-6} Pa) is between 1,000 ~ 2,000 degrees centigrade, so the melting point of the evaporation-source material must be much higher than the temperature.

(2) *Saturated vapor depression.* To reduce the evaporation-source material into the vapor-coating layer as an impurity, the saturated vapor pressure must be sufficiently low to ensure the smallest amount of self-evaporation without affecting the degree of vacuum or fouling the film.

(3) *Stable chemical properties.* Chemical reactions should not occur with raw materials at elevated temperatures. Typically, we use materials like W, Mo, Ta, or other high-temperature resistant oxides, ceramics, or graphite crucibles. Tungsten and molybdenum have a high melting point, but must be less malleable. Tantalum is the most flexible and most expensive, but easy to process.

When choosing an evaporation source, the wettability between the raw material and the evaporation-source material must be considered. In the case of infiltration, the molten-evaporation material is tiled on the surface of the evaporation source. In the absence of infiltration, the fused material will agglomerate into a sphere, which is generally considered to be a point-evaporation source. In addition, if infiltration occurs, the raw material is associated with the evaporation source and the evaporation state is stable; if it is not infiltrated, the raw material easily falls off from the evaporation source.

The resistance evaporation sources can be made into various shapes, such as boat like and silk like, as shown in Figure 4.3.

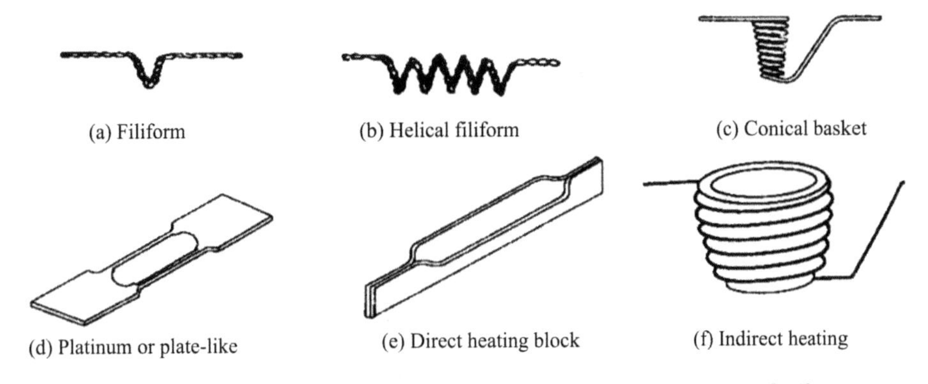

Figure 4.3 Resistance evaporation sources of various shapes[1, 2]

The line diameter of the filamentous-evaporation source is generally 0.5 ~ 1 mm. Multiple strands are often used to prevent breakage. A spiral filamentous source of evaporation is used in evaporating aluminum, because aluminum and tungsten can be wetted with each other. However, tungsten has a certain solubility in molten aluminum, so attention should be paid to it. A conical basket-shaped steaming source is generally used to evaporate bulk or filamentous sublimation materials (such as Cr) and materials that are difficult to wet with evaporation sources (such as Ag and Cu, etc.). The boat-shaped evaporation source (Figure 4.3) is often 0.05~0.15 mm in thickness. As the boat-shaped steaming source has a larger heat-dissipation surface, it consumes more power than the filamentous evaporation source. An indirect heating-type evaporation source can be formed by heating the wire outside of the ceramic crucible. The electric current heats the crucible by heating the wire. For powder evaporation materials, such evaporators are used.

4.1.2 Electron-beam evaporation source

The evaporation temperature of the resistance evaporation source is low, so the refractory metal and ceramic film cannot be evaporated. High-energy electron beams must be used as sources of heating for electron-beam evaporation. The electron beam is produced with an electron gun; the electrons emitted by the cathode in the electron gun acquire kinetic energy under the acceleration by the electric field, bombarding the raw material at the anode and vaporizing it to realize an evaporation coating.

The advantages are as follows:

(1) A high working temperature can evaporate W, Mo, Ge, SiO_2, Al_2O_3 and other high-melting-point raw materials at a high rate.

(2) The raw materials are placed in a water-cooled crucible; the container can avoid vaporization of the material and improve the purity of the coating.

(3) Surface electron-beam direct heating of raw materials, heat conduction, and thermal radiation loss are small, so the thermal efficiency is high. Of course, the structure of electron-beam heating source is complex and expensive.

E-beam evaporation sources can be divided into several types, include straight guns, E guns, and others. The straight gun is a symmetrical linear-accelerating electron gun with a power ranging from several hundred to several thousand watts. The electric-deflection coil can be used to control the electronic spot by scanning along the surface of the raw material, and it is easy to adjust and control. Its main disadvantages are large size and filament pollution. The E-type electron gun is widely used for electron-beam evaporation, and its structure is shown in Figure 4.4. The hot electron emission is accelerating the anode filament. Under the Lorenz force of the magnetic field, the electron beam is deflected by 270 degrees during bombardment heating of the raw materials.

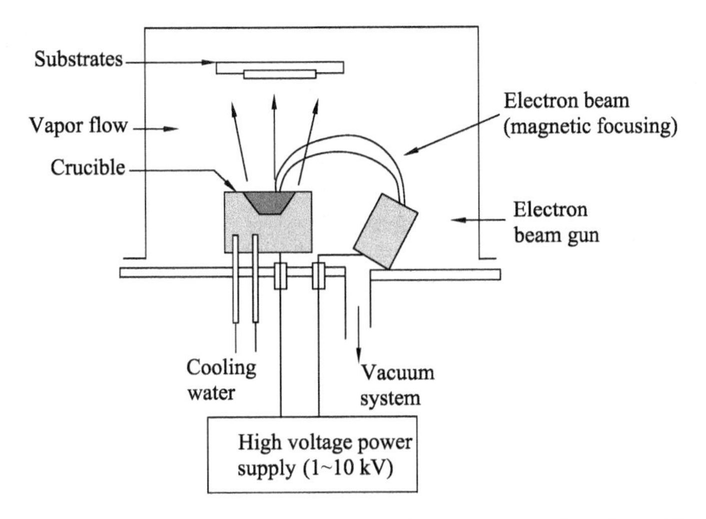

Figure 4.4 Schematic picture of the electron-beam evaporation deposition system

4.1.3 High-frequency induction-evaporation source

Figure 4.5 shows the crucible in a spiral coil. The raw material in the crucible is heated by electromagnetic induction until vaporization and evaporation are realized via the high-frequency power supply.

A high-frequency induction-evaporation source is characterized by: (1) high evaporation rate; (2) temperature uniformity and stability; and (3) relative ease of temperature control. The drawback is the need for more complex and expensive high-frequency power supplies.

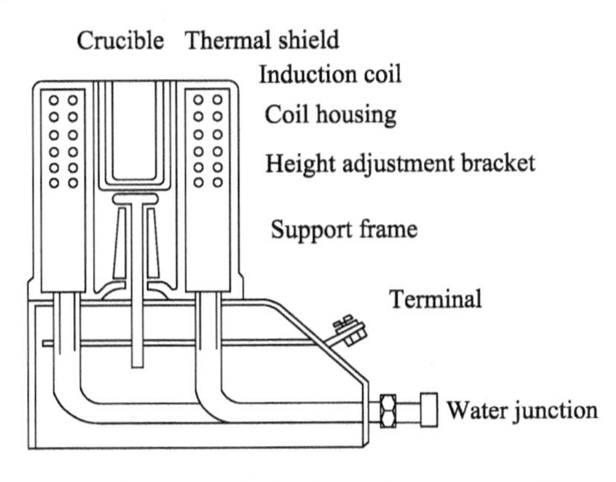

Figure 4.5 High-frequency induction of the evaporation source[1, 2]

4.1.4 Evaporation and transportation of raw materials

4.1.4.1 Evaporation rate

Assuming that the liquid-surface and gas-phase molecules are in dynamic equilibrium at the material's surface, the number of molecules evaporated from the unit area per unit time (i.e. the evaporation rate) can be expressed as

$$J_e = \frac{dN}{A \cdot dt} = \frac{\alpha_r(p_r - p_0)}{\sqrt{2\pi mkT}}, \tag{4-1}$$

where dN is the number of evaporation molecules, α_r is evaporation coefficient, A is evaporation surface area, t is the time, p_r and p_0 are the saturated vapor pressure and liquid static pressure, T is the temperature, and K is the Pohl–Seidman constant.

When $\alpha_r = 1$ and $p_0 = 0$ are obtained, the maximum evaporation rate is

$$J_m = \frac{dN}{Adt} = \frac{p_r}{\sqrt{2\pi mkT}} \tag{4-2}$$

$$\approx 2.64 \times 10^{21} p_r \left(\frac{1}{\sqrt{TM}}\right) [\mathrm{A} / (\mathrm{cm}^2 \cdot \mathrm{s})], \ (p_r \text{ in Pa})$$

where M is the molar mass of an evaporating substance.

If Formula (4-2) is multiplied by the atomic or molecular mass, the mass-evaporation rate per unit area is obtained:

$$G = mJ_m = \sqrt{\frac{m}{2\pi kT}} \cdot p_r, \tag{4-3}$$

$$\approx 4.37 \times 10^{-3} \sqrt{\frac{M}{T}} \cdot p_r [\mathrm{kg} / (\mathrm{M}^2 \cdot \mathrm{s})], \ (p_r \text{ in Pa}).$$

This expression is important for the evaporation rate; it determines the relation between the evaporation rate, vapor pressure, and temperature. The evaporation rate largely determines the temperature of the evaporation source, and the between evaporation rate and temperature are related as

$$\frac{dG}{G} = \left(2.3\frac{B}{T} - \frac{1}{2}\right)\frac{dT}{T}. \tag{4-4}$$

For metals, $2.3\frac{B}{T}$ is usually between 20~30; therefore,

$$\frac{dG}{G} = (20 \sim 30)\frac{dT}{T}. \tag{4-5}$$

Thus, a small change in the temperature of an evaporation source can significantly change the evaporation rate. Therefore, the temperature of the evaporation source must be precisely controlled during deposition.

4.1.4.2 Requirements for residual gas pressure in the coating room

Evaporation and deposition are performed under certain vacuum conditions. The residual gas molecules in the coating have important effects upon the formation and structure of the films. According to the kinetic theory of gas molecules, the molecular number per unit time per unit area under heat balance is given by

$$N_0 = \left(\frac{p}{2\pi mkT}\right)^{\frac{1}{2}}, \tag{4-6}$$

where p is the gas pressure, m is the molecular mass, and T is the gas temperature.

According to this formula, when the gas pressure $p \sim 1.3\times10^{-4}$ Pa, $T=300$ K, and adhesion coefficient $a \approx 1$, for air, N_0 is about 3.7×10^4. This shows that about 10^4 molecules of gas per second will reach the unit substrate's surface. Therefore, to obtain high-purity films, the pressure of the residual gas must be very low.

On the other hand, the evaporated material molecules move toward the substrate in the residual gas and collisions occur between particles. The average distance between two collisions is called the mean free path λ of an evaporating molecule:

$$\lambda = \frac{1}{\sqrt{2}n\pi d^2} = \frac{kT}{\sqrt{2}\pi pd^2} = \frac{3.107\times10^{-18}T}{pd^2} \text{ (p unit Pa)}. \tag{4-7}$$

In the formula, n is the residual gas density, d^2 is the collision cross section with unit $(\text{Å})^2$.

The collision probability between the evaporating molecule and the residual gas molecule can be calculated. Set N evaporation molecules after the flight distance of x; the number of non-collisions with residual gas molecules is

$$N_x = N\mathrm{e}^{-x/\lambda}; \tag{4-8}$$

the percentage of molecules colliding is

$$f = 1 - \frac{N_x}{N} = 1 - \mathrm{e}^{-x/\lambda}. \tag{4-9}$$

According to Formula (4-9), about 63% of evaporating molecules suffer collisions when the average free path is equal to the distance between the evaporation source and the substrate. If the average free path is increased by ten times, the collision probability will be reduced to around 9%. It can be seen that, only when the average free path is much larger than the source-base distance can collisions in the transport of evaporation molecules be effectively reduced. Since the average free path is determined by the pressure, the reduction of the residual gas pressure and enhancement of the vacuum is the key to reducing the collision loss of the vaporized molecules in the transport process.

4.1.5 Film-thickness distribution of the evaporation coating

In the preparation of thin films, it is generally expected that a uniform film can be obtained on the substrate. The thickness of this film depends upon the geometry of the evaporation source and the evaporation characteristics, the geometry of the substrate, the relative position of the substrate, and the source of evaporation.

The distribution of the film thickness can be calculated by the theoretical model. The model simplifies the evaporation process as follows: (1) there is no collision between the vaporized particles and the residual gas molecules; (2) each particle arriving at the substrate is condensed into a thin film. Although these assumptions do not fully conform to the actual evaporative-deposition process, they are close to the actual conditions for evaporation at lower pressures and lower substrate temperatures.

4.1.5.1 Influence of evaporation-source geometry upon the film-thickness distribution

4.1.5.1.1 Point-evaporation source

A small globular source of evaporation is called a point vapor source (referred to as point source). The evaporation characteristic of a point evaporation source is isotropic; that is, it evaporates particles in different directions at the same rate. Assuming that point source A has an evaporation rate of M g/sec, in any unit time, in any direction, the evaporation amount is $\mathrm{d}m$ in the solid angle is $\mathrm{d}\omega$, as shown in Figure 4.6.

$$\mathrm{d}m=\frac{m}{4\pi}\cdot \mathrm{d}\omega \tag{4-10}$$

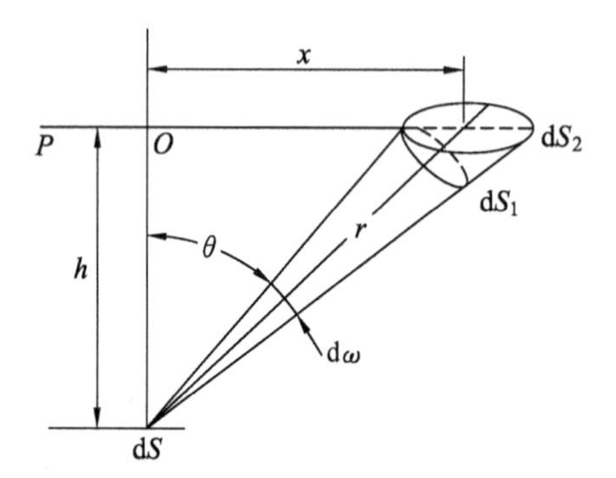

Figure 4.6 Point-evaporation source[1, 2]

For the micro-area element $\mathrm{d}S_2$ with an included angle of θ between the planar substrate and the point source, $\mathrm{d}S_1=\mathrm{d}S_2\cdot\cos\theta, \mathrm{d}S_1=r^2\cdot\mathrm{d}\omega$ are shown by the graph

$$\mathrm{d}\omega=\frac{\mathrm{d}S_2\cdot\cos\theta}{r^2}=\frac{\mathrm{d}S_2\cdot\cos\theta}{h^2+x^2}. \tag{4-11}$$

In the formula, r is the distance between the point source and the $\mathrm{d}S_2$ area element on the substrate. Therefore, the quality of the material evaporated and deposited on area $\mathrm{d}S_2$ is $\mathrm{d}m$:

$$dm = \frac{M \cos\theta}{4\pi \ r^2} dS_2. \tag{4-12}$$

Assuming that the film density is ρ, the thickness of the film deposited on dS_2 at unit time is t, and the deposited film is $t \cdot dS_2$ in volume; therefore,

$$dm = \rho \cdot t \cdot dS_2. \tag{4-13}$$

By substituting this value into Formula (4-12), the film thickness at any point on the substrate is obtained:

$$t = \frac{M}{4\pi\rho} \cdot \frac{\cos\theta}{r^2}, \tag{4-14}$$

which is a function of the coordinates (r, θ) on the substrate.

If the distance from the substrate to the point of the source plane, h, and any point on the substrate to the coordinate origin x is used to indicate the film-thickness distribution, then

$$t = \frac{Mh}{4\pi\rho r^3} = \frac{Mh}{4\pi\rho\,(h^2 + x^2)^{\frac{3}{2}}}. \tag{4-15}$$

When dS_2 is just above the point source (i.e. $\theta = 0$), $\cos\theta = 1$ and t_0 represents the film thickness at the origin; then

$$t_0 = \frac{M}{4\pi\rho h^2} \tag{4-16}$$

Obviously, t_0 is the maximum film thickness that can be obtained in the substrate plane. Thus, the relative distribution of the intimal thickness in the substrate plane is as follows:

$$\frac{t}{t_0} = \frac{1}{[1 + (x/h)^3]^{\frac{3}{2}}}. \tag{4-17}$$

4.1.5.1.2 Small-plane evaporation source

For small-plane evaporation sources (Figure 4.7), the evaporation characteristics are directional. Its evaporation in direction θ is proportional to $\cos\theta$. θ is the angle between the center line of the plane normal source of evaporation and the center line of the plane element dS_2. When the raw material evaporates from the small planar element dS_1 on the evaporation source at a rate of m grams per second, the amount of evaporation in the $d\omega$ is dm per unit time for a solid angle of θ in the direction of dS_1 in the normal plane to the small plane:

$$dm = \frac{M}{\pi} \cdot \cos\theta \cdot d\omega. \tag{4-18}$$

Here, $1/\pi$ appears because the evaporation range of small plane source is limited in the hemisphere shape space $(0 \sim \pi)$.

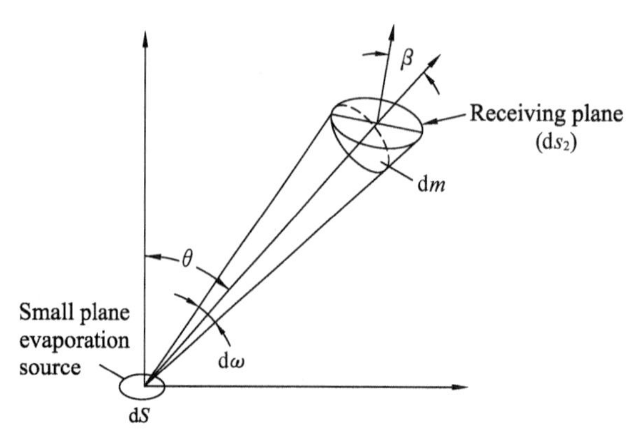

Figure 4.7 Small planar evaporation source[1, 2]

As shown in Figure 4.7, if the raw material reaches the geometry of the small plane dS_2 on the substrate with an angle β in the direction of evaporation, the amount of evaporation deposited upon the small plane per unit time (i.e. the deposition rate) can be obtained as

$$dm=\frac{M\cdot\cos\theta\cdot\cos\beta\cdot dS_2}{\pi r^2}. \tag{4-19}$$

In the same way, ρ is the specific weight of the film material and t is the film thickness. When $dm=\rho\cdot t\cdot dS_2$ is inserted into the upper form, the small-plane evaporation source can be obtained, and the film deposited at any point on the substrate has thickness

$$t=\frac{M}{\pi\rho}\cdot\frac{\cos\theta\cdot\cos\beta}{r^2}=\frac{Mh^2}{\pi\rho\left(h^2+x^2\right)^2}. \tag{4-20}$$

When dS_2 is just above the small-plane evaporation source ($\theta=0,\beta=0$), the film thickness of the point is expressed by

$$t_0=\frac{M}{\pi\rho h^2}. \tag{4-21}$$

Similarly, t_0 is the maximum film thickness obtained in the substrate plane; the film thickness in other parts of the substrate is relatively distributed, i.e. the ratio of t to t_0 is

$$\frac{t}{t_0}=\frac{1}{\left[1+(x/h)^2\right]^2}. \tag{4-22}$$

It can be seen that the thickness of the films prepared on the parallel flat substrate is uneven using the small-plane evaporation source. At the top of the plane source, the film thickness is maximal; it decreases outward.

Figure 4.8 compares the relative thickness-distribution curves of a point evaporation source and a small-plane evaporation source. Combined with the film-thickness distribution given by Formula (4-14) and Formula (4-20), we can see that the thickness distribution of the film with two kinds of source is very close. But with the same weight of raw materials and the same distance between the evaporation source and substrate, the maximum thickness of the thin film with planar evaporation source is about four times of that of point source. This can also be obtained by comparing Formula (4-16) to (4-21). The flat evaporation source saves more film material than the point source; this is because the particles evaporated from the flat evaporation source fall within the range of 0 to π, while an isotropic point source evaporates particles over 4π steradians. For a planar substrate, a planar source saves more film material than the point source.

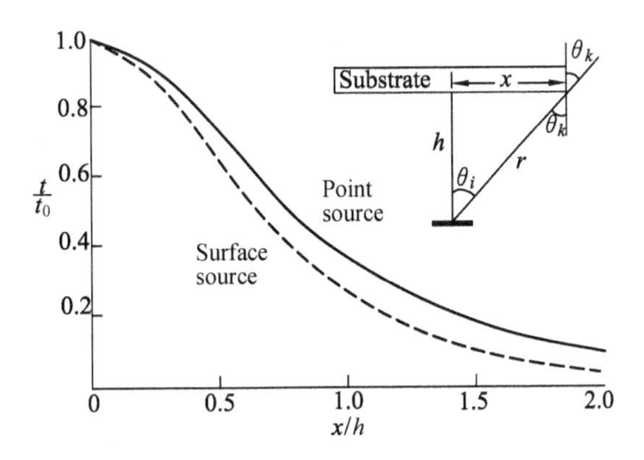

Figure 4.8 Distribution of the deposited film thickness on the substrate plane

4.1.5.2 Relative position configuration of the evaporation source and the substrate

4.1.5.2.1 Configuration of relative positions between the point source and the substrate

As shown in Figure 4.9, to obtain a uniform film thickness, an evaporation source must be configured to the center of the sphere surrounded by the substrate, such that $\cos\theta=1$ in Formula (4-21). Here, the film thickness t is constant

$$t=\frac{M}{4\pi\rho}\cdot\frac{1}{r^2}. \qquad (4\text{-}23)$$

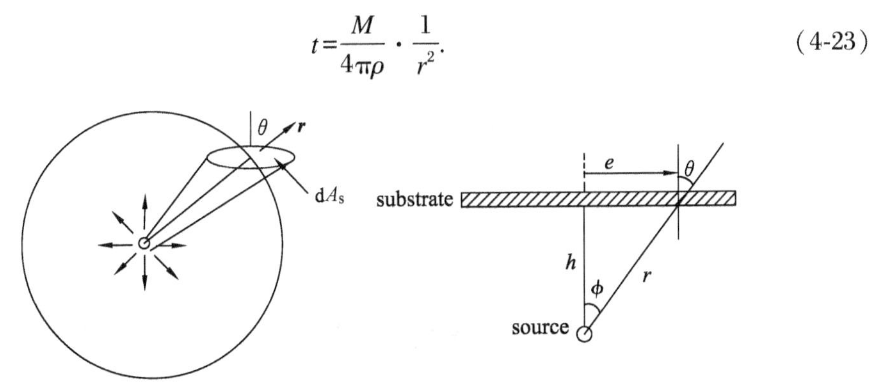

Figure 4.9 Film-thickness surface of point-evaporation source-thickness at positions l and ϕ.

Thus, only the film thickness and raw material properties such as density ρ, radius r, and quality m affect the evaporation. The theoretical results show that a spherical arrangement guarantees uniformity of the film thickness.

4.1.5.2.2 Relative positional configuration of the facet evaporation source and substrate

When the evaporation of the small plane originates from part of the spherical workpiece, the film thickness on the sphere's inner surface is uniform. This can be seen from Equation (4-20). Because when this formula is $\theta=\beta$, we have $r=2R\cos\theta$ from Figure 4.8. Its substitution type (4-20) is obtained as

$$t=\frac{M}{4\pi\rho}\cdot\frac{1}{r^2}. \tag{4-24}$$

Thus, the distribution of the film thickness t is independent of θ, such that, for a spherical workpiece with a certain radius, the film thickness of its inner surface depends only on the nature of the raw material, the r value, and the evaporation quality.

4.1.5.2.3 Location configuration of the evaporation source on a small-area substrate

If the substrate area is relatively small, the evaporation source can be directly deposited on its central line. As shown in Figure 4.10 and Figure 4.11, the substrate distance from the evaporation source, H, can be taken as $H=D\sim1.5D$, where D is the substrate-diameter dimension.

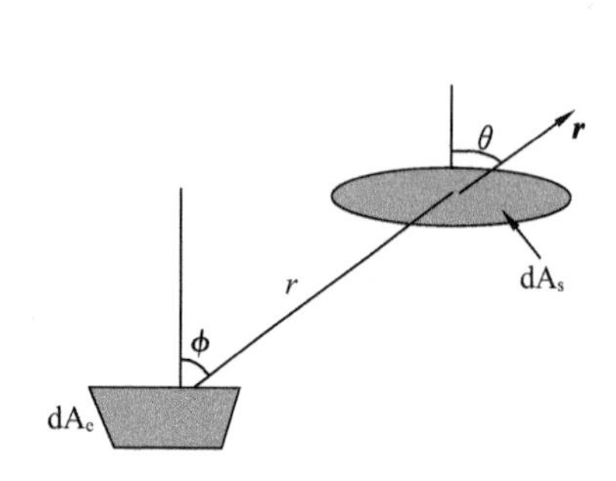

Figure 4.10 Equal-thickness surface of a small-plane evaporation source

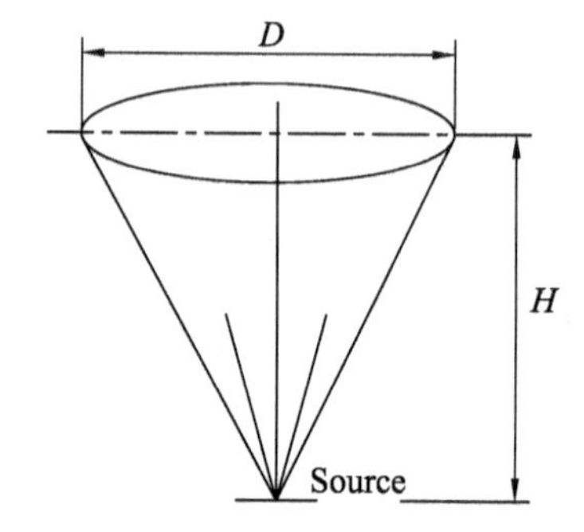

Figure 4.11 Configuration of the small-area evaporation source[1, 2]

4.1.5.2.4 Configuration of the large-area substrate and evaporation source

To obtain a uniform film thickness, in addition to using the substrate-rotation method, the use of multiple point sources instead of a single point source (or a small-plane evaporation source) is a simple and convenient method. At this point, the evaporated film thickness is distributed as follows:

$$\varepsilon=\frac{t_{max}-t_{min}}{t_0}. \tag{4-25}$$

The maximum relative deviation of film thickness falls in the range $\varepsilon-x\leqslant|1\pm1/2|$, where x is the base size. The maximum film thickness is in the range of $t_{max}-x\leqslant|1\pm1/2|$, and the minimum film thickness is in the range of $t_{min}-x\leqslant|1\pm1/2|$. The film thickness at the origin is $t_0-x=0$.

Figure 4.12 shows the uniformity of film thickness in the x direction when 4 steaming sources are used. Therefore, the location of the evaporation source and the evaporation rate obviously influence the uniformity of film thickness.

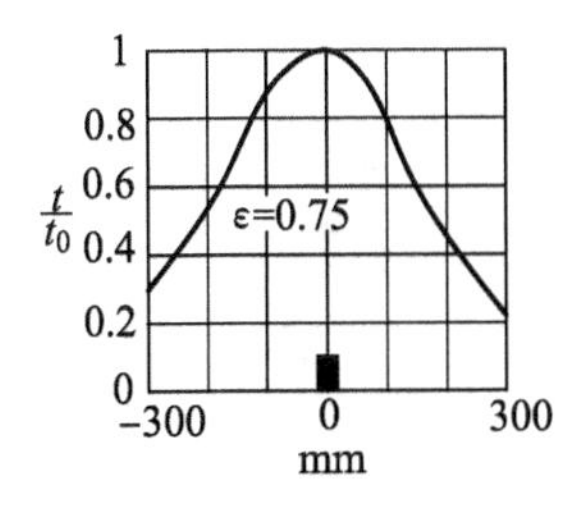

(a) One-point evaporation sources

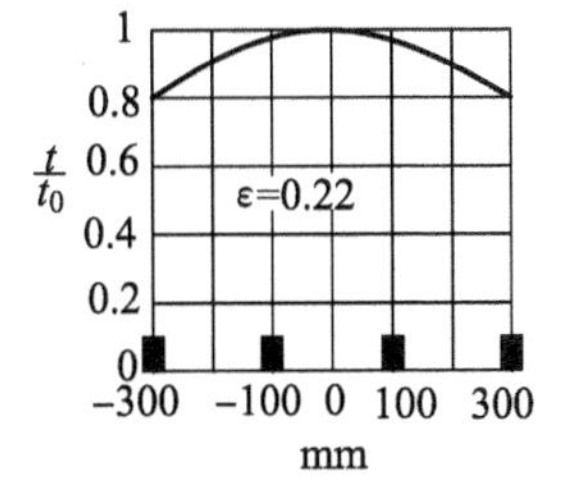

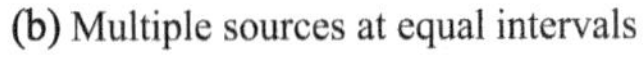
(b) Multiple sources at equal intervals

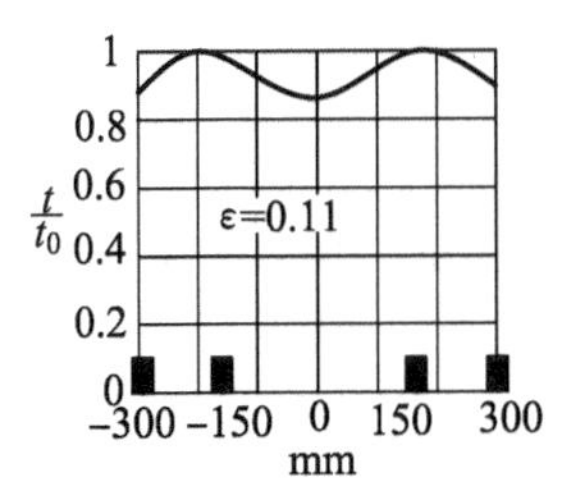

(c) Multiple sources at unequal intervals

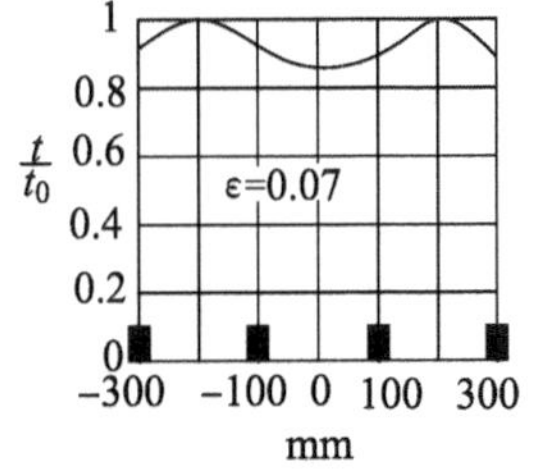

(d) Equal intervals with unequal evaporation rates

Figure 4.12 Effect of the configuration of multiple point sources upon the uniformity of film thickness t/t_0 [1, 2]

4.2 Sputtering deposition

The phenomenon whereby charged particles bombard solid surfaces and cause solid atoms (or molecules) to escape is called sputtering. The escaped particles are called sputtered particles. Charged particles used for bombardment can be electrons, ions, or neutral particles. Because the ion mass is much larger than that of electrons, most ions are used as bombarding particles (also known as incident particles). Sputtering particles are deposited upon the substrate surface to form a thin film via a process called sputter deposition (Figure 4.13). This has been widely used to prepare various films, including metals, alloys, semiconductors, oxides, carbides, nitrides, and super-conductors.

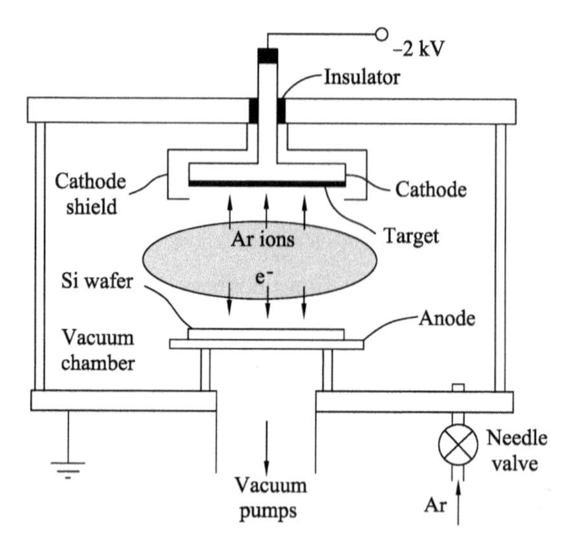

Figure 4.13 Schematic picture of a sputter-deposition system[15]

4.2.1 Sputtering characteristics

Understanding the sputtering characteristics is important for understanding the sputtering process. Sputtering parameters, threshold, etc. are the main parameters for characterizing sputtering characteristics.

The sputtering rate is the most important physical quantity describing the sputtering characteristics. In sputtering-deposition devices, the sputtering target is usually placed at the cathode and hence referred to as the target cathode. The sputtering rate means that when the particle hits the target cathode, the average number of atoms produced by each particle from the cathode, also known as the sputtering yield or coefficient, is expressed by S. The sputtering rate is related to the type, energy, angle, and surface state of the incident particles.

4.2.1.1 Target material

The sputtering rate is related to the location of the target element in the periodic table. The experimental results show that the sputtering rate changes periodically—in general, it increases with its target's atomic number. The sputtering rates of copper, silver, and gold are the highest, and those of carbon, silicon, titanium, vanadium, zirconium, niobium, tantalum, and tungsten are smaller. Analysis of the atomic structure shows that the sputtering rate is related to the filling degree of the $3d$, $4d$, and $5d$ shells of atoms.

4.2.1.2 Incident particle energy

Sputtering occurs when the incident particle energy is higher than a critical value. This critical value is called the sputtering threshold. When bombarding copper with Ar^+ ions, the typical relationship between the ion energy and the sputtering rate is shown in Figure 4.14(a). This curve can be divided into three parts: Part I ($E<70$ eV) is the low-energy region, in which no sputtering occurs. Part II (70 ~ 10 keV) is the region where the sputtering rate increases with ion energy; most of the ion energy deposited by sputtering falls within this range. Part III (above 30 keV) is the region where the sputtering rate decreases along with ion energy. Here, high-energy bombarding ions are introduced into the crystal lattice, resulting in fewer sputtering particles.

4.2.1.3 Kinds of incident particles

In general, the higher the atomic mass of the incident particle, the higher the sputtering rate. The sputtering rate also varies periodically with the atomic number as shown in Figure 4.14(b). Any element with a filled electron shell has the highest sputtering rate. Therefore, the sputtering rate of the inert gas is the highest; thus, argon is often used as a sputtering gas. Another advantage of using inert gas is that chemical reactions with the target can be avoided.

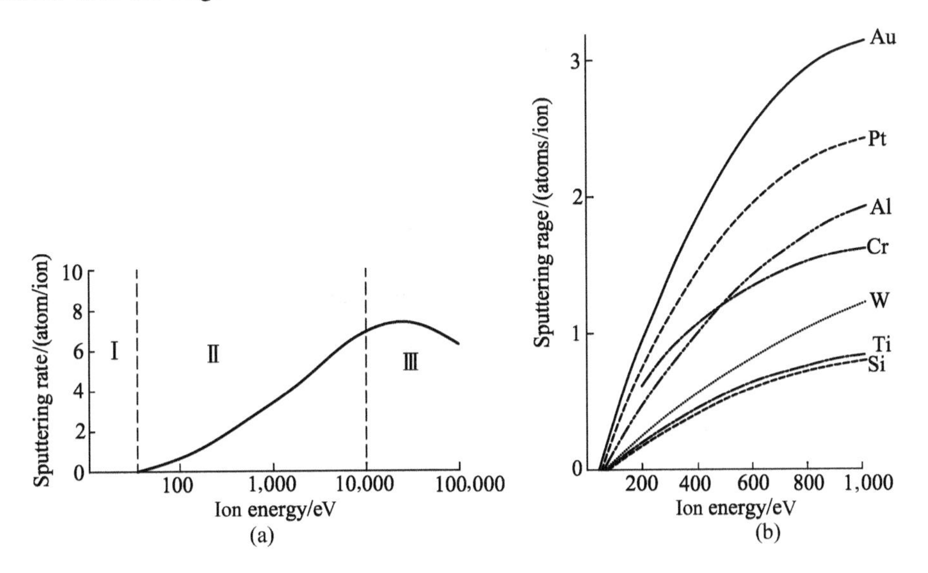

Figure 4.14 Relationship between ion energy and sputtering rate of Ar^+ ions for (a) copper, and (b) other material: Au, Pt, Al, Cr, W, Ti and Si[13]

4.2.1.4 The incidence angle of the incident particle

The incidence angle falls between the incident direction of the Ar^+ and the normal surface of the target. Figure 4.15 shows the relationship between the sputtering rate and the incident angle of Ar^+ to several metals. It can be seen that the relative sputtering rate in the region 0~60° basically obeys $1/\cos\theta$. The sputtering rates of $S(\theta)$ and $S(0)$ have incident angles of θ and vertical incidence, respectively. When the incident angle is 60° ~ 80°, the sputtering rate is the highest. When the incident angle further increases, the sputtering rate decreases rapidly, and the sputtering rate is zero when equal to 90°. Therefore, there is an optimal incidence angle S for the maximum sputtering rate θ_m.

In addition to the above factors, the sputtering rate is also related to the target structure, the crystal orientation of the target, the surface morphology, and the sputtering pressure.

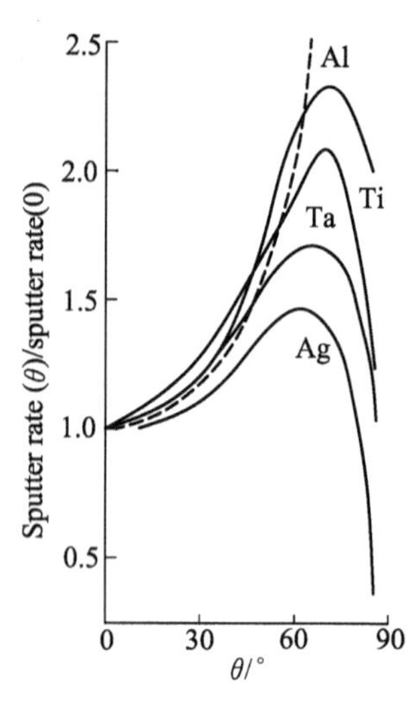

Figure 4.15 Relationship between the incident angle of Ar^+ and the sputtering rates of several metals[13]

4.2.2 Sputtering-deposition process

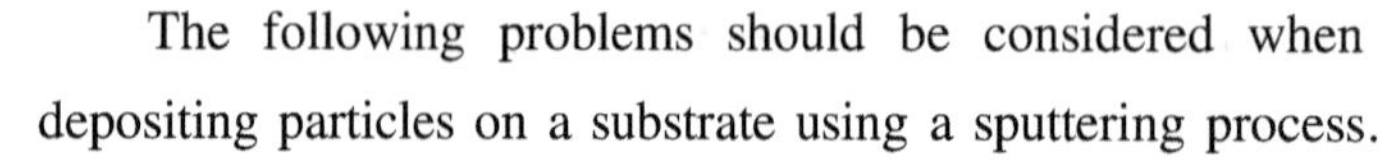

The following problems should be considered when depositing particles on a substrate using a sputtering process.

4.2.2.1 Deposition rate

Deposition rate Q is the thickness of the deposited material on a substrate per unit time. The rate is proportional to the rate S of sputtering,

$$Q = CIS \tag{4-26}$$

In the formula, C is the characteristic constant related to the sputtering device, I is the ion current, and S is the sputtering rate.

4.2.2.2 Deposition pressure

To improve the purity of the deposited films, the amount of residual gas entering into the film must be reduced. Typically, a few percent of the sputtering gas molecules are deposited onto the film, especially when the substrate is biased. If the vacuum-chamber volume is V, the residual gas partial pressure is p_r, the argon partial pressure is p_{Ar}, the residual gas-flow rate is Q_r, and the argon-flow rate is Q_{Ar}, then we have

$$Q_r = p_r V,$$
$$Q_{Ar} = p_{Ar} V,$$
$$p_r = p_{Ar} Q_r / Q_{Ar}. \tag{4-27}$$

Therefore, two effective measures are adopted to increase the vacuum level of the bottom and increase the argon-flow rate. Generally speaking, the background vacuum should be 10^{-3} to 10^{-5} Pa, and the argon pressure should be about several Pa, which is more appropriate.

4.2.2.3 Process conditions of sputtering deposition

Many factors influence sputtering deposition, including sputtering voltage and substrate potential (ground, suspension, or bias); these have a major influence upon the film characteristics. The sputtering voltage not only affects the deposition rate, but also the structure of the films. If there is an alternating bias on the substrate, with a positive value attracting electrons and a negative attracting ions, this can not only purify the substrate surface and enhance the film's adhesion, but also change the crystalline structure of the deposited films. In addition, the substrate temperature directly influences the growth and properties of the films.

4.2.3 Plasma-sputtering deposition

Sputtering deposited ions are produced in a gas-discharge plasma. Depending on how the discharge is formed, sputtering coating can be divided into direct current (DC) sputtering, radio frequency (RF) sputtering, reactive sputtering, and magnetron sputtering.

4.2.3.1 DC two-stage sputtering

The DC sputtering power supply can form a DC-sputtering-deposition film system. A DC two-stage sputtering system is formed by placing the target on the cathode and placing the substrate on the grounded electrode (anode). A schematic diagram of the structure is shown in Figure 4.16. At work, the vacuum chamber is first pumped into a high vacuum (e.g., 10^{-3} Pa); then, argon is injected to maintain the pressure at 1~10 Pa. At this time, the power supply is connected with an abnormal glow discharge between the cathode and the anode. Gas ionization forms the plasma. The positively charged argon ion is accelerated by the electric field, bombarding the cathode target, such that the target is sputtered. A common DC-sputtering power supply is typically 500 W~1 kW; the rated current is 1 A; and the voltage is 0~1 kV.

The DC-diode sputtering structure is simple, but it has the following shortcomings:

(1) The discharge current varies with voltage and pressure and the sputtering parameters are not easily controlled;

(2) The deposition rate is low;

(3) The target must be a conductor.

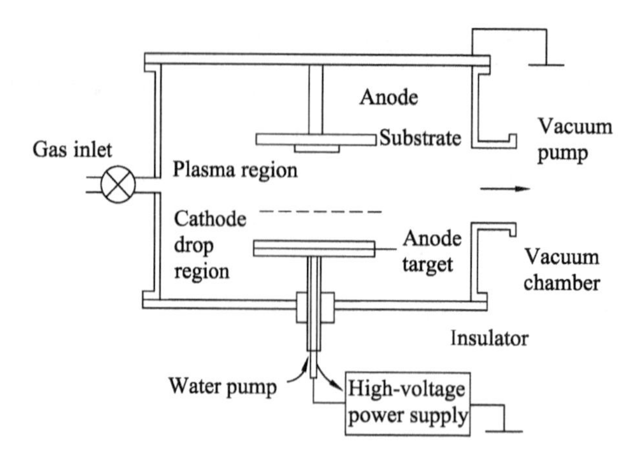

Figure 4.16 DC-diode sputtering system[13]

4.2.3.2 RF sputtering

An AC-sputtering system uses an AC (rather than a DC) power supply. Because the frequency of a common AC power supply is in radio frequency (RF) range (e.g., 13.56 MHz), this is called RF sputtering, as shown in Figure 4.17.

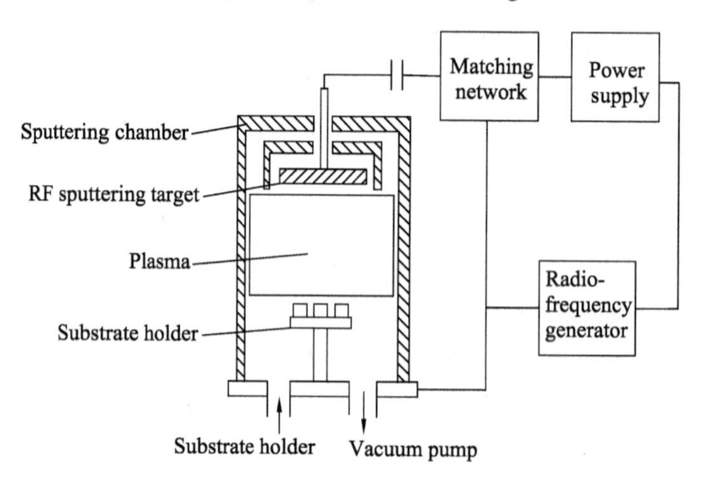

Figure 4.17 RF sputtering system[13]

If an insulating material is used in a DC-sputtering device, the target ions will accumulate on the target surface, making it positively charged. The target potential is raised such that the electric field between the electrodes becomes smaller until the glow discharge is extinguished and the sputtering is stopped. Therefore, the DC-sputtering device cannot be used for sputter deposition of insulating dielectric films. To sputter-deposit insulating materials, DC power supplies must be replaced with AC ones. Due to the positive and negative polarity of the alternating-current power source, when the sputtering target is in the positive half cycle, the electrons flow to the target surface, neutralize the positive charge accumulated on the target surface, and accumulate electrons such that the target surface is negatively biased. The positive-ion-bombardment target is attracted at the negative half of the RF voltage to achieve sputtering. It is important to note that an RF sputtering device may also sputter the conductor target due to negative bias there. The RF sputtering device is shown in Figure 4.17, and the power supply is composed of an RF generator and a matching network. The matching network is used to adjust the input impedance to match the output impedance of the RF power supply, so as to achieve maximal power-transmission efficiency.

In an RF sputtering device, the electrons in the plasma easily absorb energy in the RF field and oscillate in the electric field. Hence, electrons collide with working gas molecules and ionize them, increasing the chance of producing ions. Therefore, the breakdown voltage, discharge voltage, and the working air pressure are reduced remarkably.

4.2.3.3 Magnetron sputtering

To perform high-speed sputtering at low pressure, the ionization rate of gas must be effectively improved. By introducing a magnetic field on the cathode surface of the target, the sputtering density can be increased using a magnetic field to restrain the charged particles and increase the sputtering rate. This is called *magnetron sputtering*.

The principle behind magnetron sputtering is shown in Figure 4.18. The electrons collide with argon atoms in the process of flying to the substrate in the presence of an electric field and ionize them to produce Ar^+ and new electrons. The new electron flies toward the substrate. Ar^+ ions fly toward the cathode in the electric field, and bombard target surface with high energy. In this case, the sputtering of target occurs. When sputtering particles, a neutral target atom or molecule is deposited onto a substrate to form a thin film. The resulting two electrons are affected by electric and magnetic fields, resulting in E (electric)×B (magnetic) pointing drift (referred to as "$E \times B$" drift), with a trajectory similar to a cycloid. In the case of an annular magnetic field, the electron undergoes circular motion near the target surface in an approximately cycloidal form. Its path of motion is not only very long, but it is also confined in the plasma region near the target surface. Additionally, a large number of Ar^+ ions are ionized in the region to bombard the target, thus achieving a high deposition rate. As the number of collisions increases, the energy of the two electrons is depleted as they gradually move away from the target surface, before finally being deposited on the substrate in the electric field E. Because the electron energy is very low, the energy transferred to the substrate is very small, resulting in a lower substrate-temperature rise.

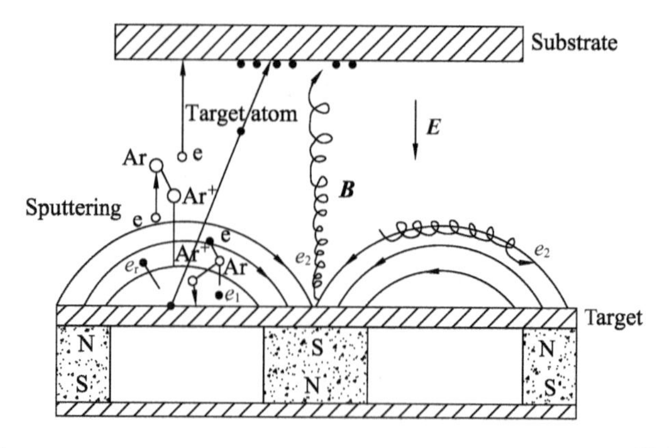

Figure 4.18 Principle behind magnetron sputtering device[1,2]

To summarize, the basic principle of magnetron sputtering is to use a magnetic field to bind electrons, prolong the existence of electrons, and increase the ionization probability of electrons in the working gas. It can also utilize the electron energy to produce higher-density positive ions, and then increase the sputtering rate of positive-ion bombardment targets. At the same time, electrons bound by an orthogonal electromagnetic field can only be deposited onto the substrate when the energy is used up. Therefore, magnetron sputtering has two characteristics: low temperature and high speed.

There are three main types of magnetron-sputtering sources, as shown in Figure 4.19. The first to be developed was a cylindrical magnetron-sputtering source, as shown in Figure 4.19(a) and 4.19(b). Its structure is relatively simple, suitable for sputtering deposition of large-area films, and widely used in industry. The third type is the planar magnetron-sputtering source, as shown in Figure 4.19(c) and 4.20. Made of small targets, it is suitable for valuable targets. A rectangular fit is made into a large target. The planar magnetron-sputtering source has the advantages of a simple structure, versatility, and wide application. The fourth type is the sputtering gun (S gun). As shown in Figure 4.19(d), the S gun has a more complicated structure and is commonly used in conjunction with planetary clamps. The S gun features low temperature; it not only has high rate of magnetron sputtering, due to its special target structure and cooling method, but also offers high efficiency, a good film-thickness distribution, target power density, and easy target replacement.

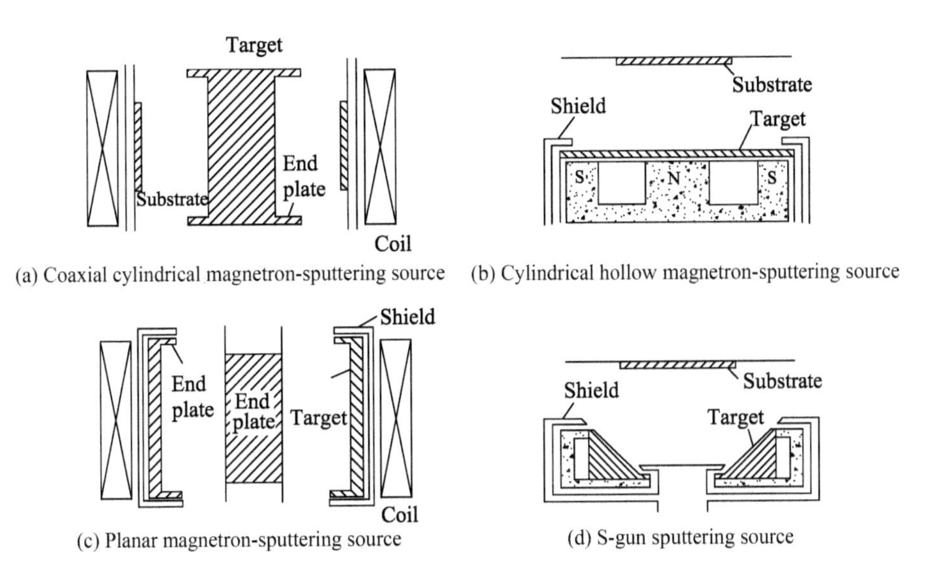

(a) Coaxial cylindrical magnetron-sputtering source (b) Cylindrical hollow magnetron-sputtering source

(c) Planar magnetron-sputtering source (d) S-gun sputtering source

Figure 4.19 Different kinds of magnetron-sputtering sources[1, 2, 13]

In general, the operating parameters of a planar magnetron-sputtering source are: sputtering voltage 300 ~ 800 V, current density 4 ~ 50 mA/cm^2, argon pressure 0.13 ~ 1.3 Pa, power density 1 ~ 36 W/cm^2, substrate-target distance 4 ~ 10 cm. Magnetron sputtering not only offers a high sputtering rate, but also prevents two-electron bombardment while sputtering the metal to keep the substrate cold. This is important for the use of single crystals and plastic substrates. A magnetron-sputtering power supply can be DC or RF, such that it can prepare various materials. However, magnetron sputtering has two problems: first, it is difficult to sputter the magnetic material target because the magnetic flux is shortened by the magnetic target; second, the sputtering and etching of the target material is uneven and the utilization rate is low.

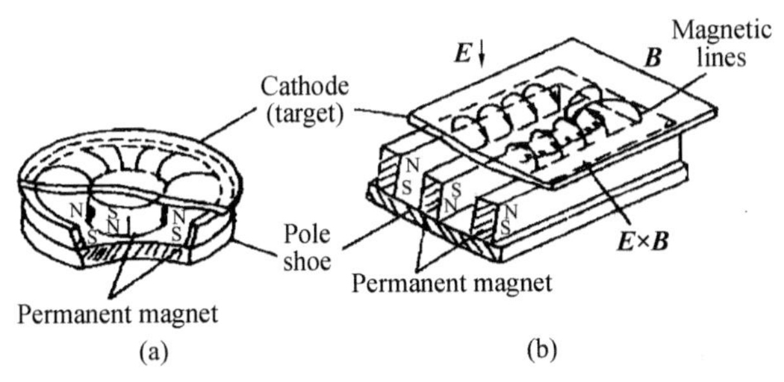

Figure 4.20 Planar magnetron-sputtering source: (a) Round and (b) Rectangle[1, 2, 13]

4.2.3.4 Enhanced magnetron sputtering

4.2.3.4.1 The sputtering of the target

For Fe, Ni, Fe_2O_3, permalloy, and other magnetic materials, it is very difficult to adopt these magnetron-sputtering methods to achieve low-temperature and high-speed sputtering deposition. This is because the magnetoresistance of a target made of such material is very low, and the applied magnetic field almost completely passes through the target, and rarely leaks. The parallel magnetic field cannot be confined to the target surface to restrict the electrons.

A target magnetron-sputtering device has been developed to deposit magnetic films. The sputtering principle of the opposite target is shown in Figure 4.21. The two targets are placed relative to each other, and the magnetic field is perpendicular to the target surface but parallel to the electric field. After two electrons fly out of the target surface, the electric field is accelerated by the cathode-drop region perpendicular to the target. The electrons move in the direction of the anode and are acted upon by the magnetic field; however, since both targets have higher negative bias, the electrons bounce back and forth between the two electrodes. The length of the electronic movement is greatly prolonged, and the collision-ionization rate with argon is increased, thereby greatly improving the density of argon ions required for sputtering and improving the deposition rate. The sputtering target has the characteristics of a high sputtering rate, a low substrate temperature, and deposition of a magnetic thin film.

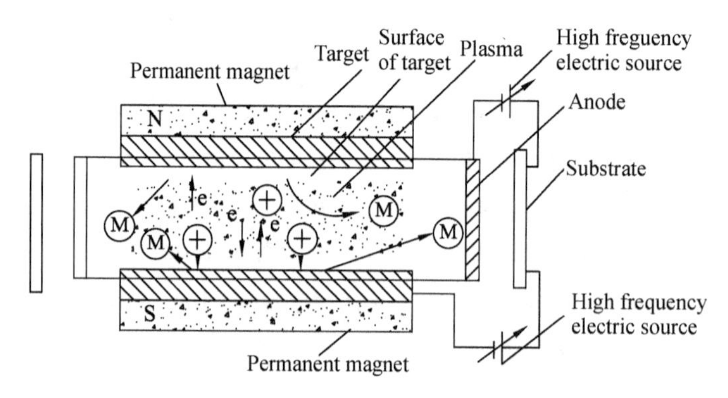

Figure 4.21 Facing target-sputtering device[13]

4.2.3.4.2 Unbalanced magnetron sputtering

If the plasma density is increased near the anode (substrate) to activate the reaction gas, more particles bombarding the cathode target will be produced, thereby increasing the sputtering rate. To achieve this, the magnetic field can be increased at the target, such that two electrons from the cathode enter the plasma region to participate in ionization, which is called unbalanced magnetron sputtering. Figure 4.22 shows the design of three different magnetic-field structures.

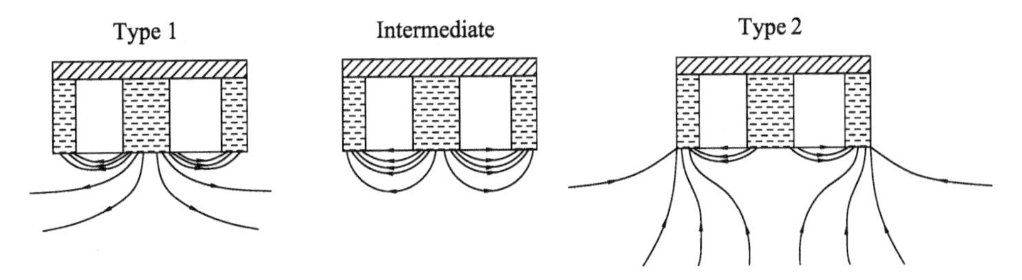

Figure 4.22 Magnetic-field structure of non-equilibrium sputtering[13]

4.3 Ion-beam deposition

The sputtering target can be used to prepare the film using a charged ion beam. Since the direction, energy, and flux of the incident ion can be controlled independently, and the growth of the film structure can be controlled precisely. One recently developed technique for doing so is double-ion-beam deposition.

4.3.1 The basic process of double-ion-beam deposition

The principle of ion-beam deposition on thin film is shown in Figure 4.23. The ion beam is produced by an ion source. In the dual-ion-beam deposition system, the first beam is the inert-gas discharge-ion source (Ion source 1) producing Ar^+ and Xr^+ which bombard the target. In this case, the sputtering occurs. The sputtered particles are deposited on the substrate, becoming a thin film. The second is a reactive ion beam (Ion source 2) that is directly aligned with the substrate, so that it dynamically irradiates the growing film. The structure and properties of the films are controlled and changed by bombardment, reaction, or intercalation.

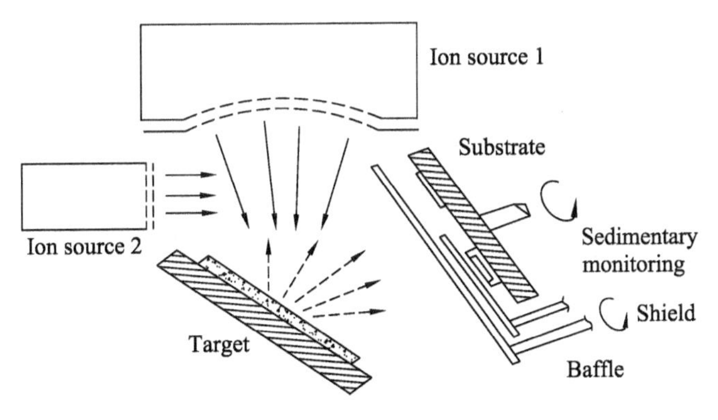

Figure 4.23 Schematic picture of ion-beam sputtering device[13]

Ion-beam-deposition technology allows multiple materials to be deposited. Although the phenomena involved in this technique are complicated, by suitably choosing the target and ion-beam energy, one can produce all kinds of metal, oxides, nitrides, and other compound films, especially multi-component metal oxides, inlay materials, functionally graded materials, and super-hard material films.

4.3.2 Ion source

An ion source is the core of ion-beam deposition system. Three kinds of ion sources are commonly used: a Kauffman electron-bombardment ion source, a high-frequency-discharge ion source, and an electron-cyclotron-resonance ion source. Among these, the Kaufman ion source is most widely used. The hot filament emits electrons in a cylindrical discharge chamber. The hollow cathode is placed on the horizontal axis of the discharge chamber and surrounded by an anode outside the cathode. At work, the gas is fed into the discharge chamber, and electrons are accelerated by an electric field between the cathode and the anode to obtain energy. Collision ionization of gas forms plasma. To effectively restrain the plasma, an axial magnetic field is applied in the discharge chamber. Ions in the plasma are accelerated by small holes in the accelerating electrode and accelerated to form an ion beam. At present, an ion source with a beam width of 1 meter has been developed. The current density reaches several mA/cm^2, and the beam energy is several thousand eV. The divergence angle is a few degrees.

4.3.3 Ion-beam modification of thin film materials

Ion-beam bombardment can also be used to improve the performance of the film, mainly by

(1) Enhancing surface-adsorbed atoms;

(2) The early steps of simulation of thin film formation, such as nucleation and growth;

(3) Promoting the preferential orientation of epitaxial growth;

(4) Lowering substrate temperature;

(5) Amorphous crystallization of amorphous thin films;

(6) Enhancing the adhesion between the film and the substrate;

(7) Improving the film stress;

(8) Simulating the thin-film-adsorption effect and film-surface reaction.

4.4 Pulsed-laser deposition

Pulsed-laser deposition is a new thin film-preparation technique using high-energy laser beams to deposit films onto material surfaces. A high-power excimer laser can be used as a laser source. A high-energy laser beam enters the vacuum chamber through the window and is focused by a prism or concave mirror. The laser beam's power density after focusing is very high, potentially exceeding 10^6 W/cm^2. Raw materials are irradiated to heat and vaporize them.

Figure 4.24 is a schematic diagram of a pulsed-laser-deposition (PLD) film device. Most non-metallic materials strongly absorb 200 ~ 400 nm UV light. The shorter the wavelength, the greater the absorption coefficient and the deeper the penetration depth. The lasers employed in PLD technology is are mainly solid-state Nd^{3+}: YAG (1,064 nm) and gas-excimer laser. Using the solid-state Nd^{3+}: YAG (1,064 nm) laser, the energy of each pulse can reach ~2 J, and the repetition rate of each pulse is ~30 Hz. Radiation of 1,064 nm can produce 355 nm and 266 nm outputs through frequency doubling and mixing. Gas-excimer lasers mainly include ArF (193 nm), KrF (248 nm), and XeCl (308 nm) types. A commercial gas-excimer laser has a pulse energy of 500 mJ and a pulse-repetition rate of several hundred Hz.

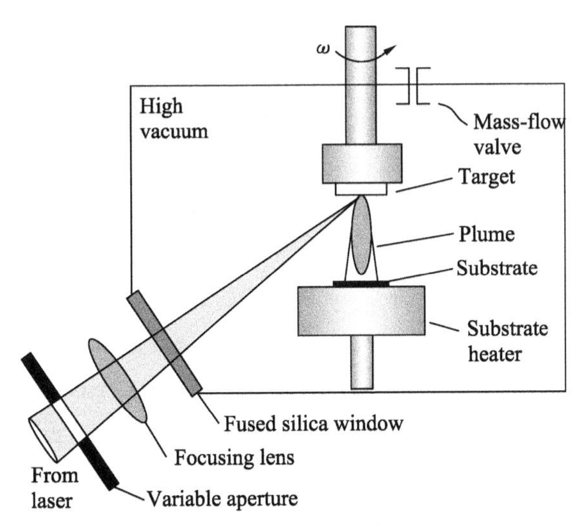

Figure 4.24 Schematic picture of the pulsed-laser-deposition device[15]

The energy of the laser beam will cause excitation of the target atom and ablation of the target surface. The plume consists of charged neutral atoms, molecules, ions, electrons, clusters, microscale particles, and molten droplets mixed in front of the target.

The body has a highly oriented pinnate accumulation and its dispersion angle is $\cos^n \phi$ ($8<n<12$). When the accumulation reaches the substrate, it condenses there to form a thin film. Gases such as O_2 and N_2 are usually introduced into the vacuum chamber to enhance the surface reaction or to maintain the film's stoichiometric ratio. When a single film is deposited, a uniform multi-unit target is usually required. However, for layered structures, many sources must be evaporated by a laser beam; this allows a single laser and beam splitter to send two or more lasers simultaneously. Or else, the single laser beam is thus focused upon different component targets on the rotating-sample shelf. The window material is an important part of the PLD system and must be transparent to visible and UV light at the same time. MgF_2, CaF_2, and UV-grade quartz can be used as window materials.

Pulsed-laser deposition has the following advantages: (1) Laser heating can achieve high temperature, evaporation can be attained for any high-melting-point materials, and the evaporation rate is high. (2) The instrument adopts non-contact heating, and the laser can be installed in the vacuum chamber. Thus, it is completely avoided as a source of pollution and the vacuum chamber can be simplified, making it very suitable for the preparation of high-purity films in ultra-high vacuum. (3) Laser heating can be used to cause "flash evaporation" of certain compounds or alloys, thereby ensuring chemical-film components or preventing the decomposition ratio; the material gasification time is too short for the surrounding materials to reach evaporation temperature, so laser-evaporation fractionation does not readily occur. Therefore, it is a good method for depositing complex films such as dielectric, semiconductor, or inorganic compounds. However, pulsed-laser deposition equipment is more expensive. Moreover, because the target's surface temperature is very high, the evaporation particles (atoms, molecules, clusters, etc.) are easy to ionize, which will have some influence upon the structure and properties of the film.

In fact, all kinds of films prepared by PVD technology can be prepared by pulsed-laser technology, including $BaTiO_3$, $SrTiO_3$, ZnS, high-T_c oxide superconducting films and other compounds, films, and diamond-like carbon films. Note that it is difficult to prepare the ceramic film at a stoichiometric ratio because other deposition technologies need react with the precursor gas. The pulsed-laser method directly adopts the stoichiometric ratio to meet the needs of sintered ceramics as target, so the pulsed-laser method is especially suitable for the preparation of ceramic film.

4.5 Molecular-beam epitaxy

Molecular-beam epitaxy (MBE) is a technique that has mainly been used in the semiconductor industry to produce thin films of compound semiconductors (e.g., GaAs, InP), as used in the fabrication of LEDs, laser diodes, etc. These inorganic semiconductors are ceramics, so it should not be surprising that the technique can also be used to grow other ceramic thin films (e.g., the high-temperature superconductor $YBa_2Cu_3O_7$). In fact, MBE is ideal for ceramics with layered structures, because it allows precise sequential deposition of single monolayers. An MBE system is shown in Figure 4.25.

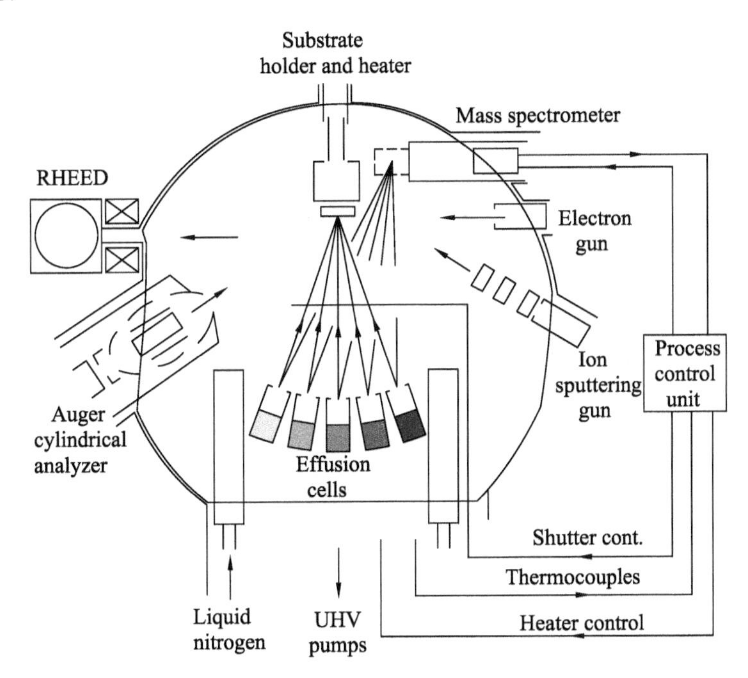

Figure 4.25 Schematic picture of the molecular-beam-epitaxy deposition system[15]

The materials to be deposited are usually evaporated from separate furnaces called Knudsen effusion cells, which are bottle-shaped crucibles with a narrow neck. A resistance heater wound around the cell provides the heat necessary to evaporate the material. As in the case of conventional evaporation, the source may be solid or liquid depending upon its vapor pressure. The rate of deposition of each species is determined by the vapor pressure above the source, and strongly depends on temperature. The temperature of each of the Knudsen effusion cell therefore controls the flux of atoms reaching the substrate. MBE of semiconductors requires the use of an ultra-high-vacuum

(UHV) chamber (background pressure: 10^{-10} ~ 10^{-8} Torr). For oxide ceramics, background pressures of 10^{-4} Torr are more common. The high-vacuum requirement for MBE presents a problem for the growth of many multicomponent oxides (e.g., high-temperature superconductors) because most of these compounds require oxygen pressures much higher than this to form. This limitation has been overcome using highly oxidizing gases (e.g., NO_2, atomic O, or O_3) near the surface of the growing film while the background pressure is kept as low as possible.

In addition to the requirements of high-vacuum or UHV environments, the other features that limit the use of MBE are:

- The equipment is expensive (> $ 1 million), so the added value must be high.
- Deposition rates are low: <1 μm/h is typical.

Exercises

1. What's physical vapor deposition?
2. How is the physical vapor deposition classified?
3. What's the working mechanism of thermal evaporation deposition?
4. How is the thickness of the thin film made by thermal evaporation calculated?
5. What's the working mechanism of electron beam thermal evaporation deposition?
6. What's the working mechanism of sputtering deposition?
7. What's the working mechanism of magnetron sputtering deposition?
8. What's the working mechanism of pulsed laser deposition?
9. What's the working mechanism of Molecular-beam epitaxy deposition?
10. Why is it often desirable to form thin films at the lowest possible substrate temperature?
11. What technique would you use to produce a 100 nm thin film of AlN on silicon? Explain why you choose your technique.
12. What technique would you use to produce a 5 nm thin film of $BaTiO_3$ on MgO? Explain why you choose your technique.
13. Pulsed laser deposition is our favorite technique for growing oxide films. What is the largest substrate currently being used? What are the thickness limitations?
14. Which of the techniques described in this chapter is most suitable for producing thin films on large substrates? What is the largest substrate that can be coated?

Chapter 5 Chemical-vapor Deposition

Chemical-vapor deposition (CVD) uses heating, plasma, lasers, and other means to stimulate a gas to produce chemical reactions. By this method, a wide range of films can be prepared, including metals and dielectrics, wear-resistant coatings, high-temperature protective coatings, and so on. The deposition rates of films prepared by CVD are high, the films have good uniformity, excellent step coverage, and are suitable for coating on complex substrates.

CVD techniques can be classified according to how chemical reactions are stimulated:

(1) Hot CVD. This is a method of growing a film by heating a substrate or a wall to promote a chemical reaction.

(2) Plasma CVD. This method uses plasma to enhance the chemical activity of the reacting gas, thereby promoting the film growth. The plasma CVD can be subdivided into direct current (DC-CVD), radio-frequency (RF-CVD) and microwave-plasma (MW-CVD) types according to the difference of the plasma's electric-field frequency. According to the size of the plasma density, CVD can be divided into high-density and low-density plasma CVD.

(3) Light CVD. The method uses light irradiation energy to promote decomposition of gaseous matter and to react to the growth of the film.

5.1 Thermochemical-vapor deposition

There are many types of CVD reactors, but all must meet the following conditions: (1) At the deposition temperature, the reactants must have a high enough saturated vapor pressure, and they are induced into the reaction chamber with a proper flow rate.

(2) Among the products of chemical reaction, except the solid state film, other reaction products should be volatile, and can be removed. (3) The vapor pressure of the substrate material at the deposition temperature must be sufficiently low, so that it will not be evaporated. (4) The reaction-chamber wall temperature must be sufficiently low with no pollution. Typical systems are shown in Figure 5.1, including intake systems, reaction chambers, exhaust and exhaust-treatment systems, heaters, etc. Such systems usually open under normal pressure and are easy to load and unload.

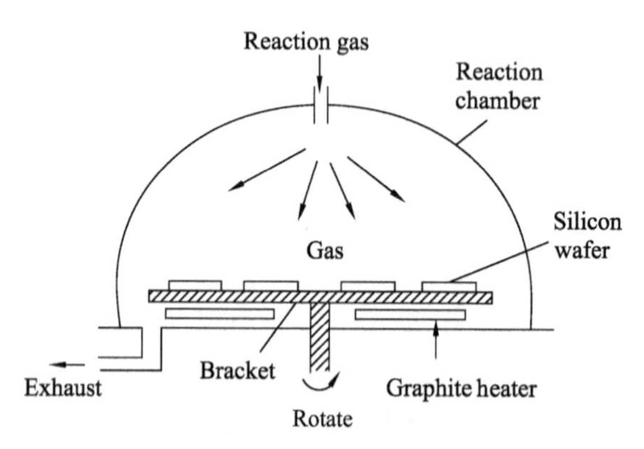

Figure 5.1 Vertical chemical-vapor deposition device[1, 13]

The raw material can be gas, or if a liquid or solid material is used, it needs to be heated to produce steam, and then carried by the gas into the furnace. Inductive heating is used in the deposition area, and the reactor wall is heated to prevent reaction. The device often operates at pressures lower than that of the atmosphere. Due to the high diffusion coefficient of gas under low pressure, the gaseous reaction agent and by-product quality increase along with the film growth rate, and a thin film of uniform thickness forms, improving the production efficiency.

Figure 5.2 is a schematic diagram of a horizontal low voltage CVD device. Single-crystal silicon and polycrystalline silicon films, silicon-nitride films, and Ⅲ ~ V group films can be prepared by a low-voltage CVD method, as well as silicon nitride, silicon dioxide, and aluminum oxide films. This method can also be used to manufacture VLSI.

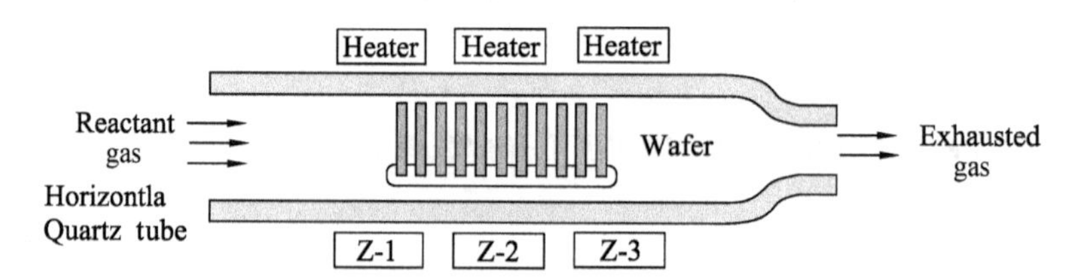

Figure 5.2 Schematic picture of horizontal chemical-vapor deposition equipment

5.2 Plasma-enhanced chemical-vapor deposition

The deposition temperature of hot CVD is higher. Although the deposition temperatures of a few films are below 500 ℃, most reaction temperatures must be 500 ℃ ~ 1,000 ℃. The high temperature causes the substrate to deform and its material to become reactive. In recent years, plasma chemical-vapor deposition has been rapidly developed and widely used to reduce deposition temperature and enhance chemical reactions.

Plasma chemical-vapor deposition activates CVD using plasma formed by glow discharges, as shown in Figure 5.3. The electron energy in the glow-discharge plasma is about 1 ~ 10 eV, equivalent to a temperature of $10^4 \sim 10^5$ K.

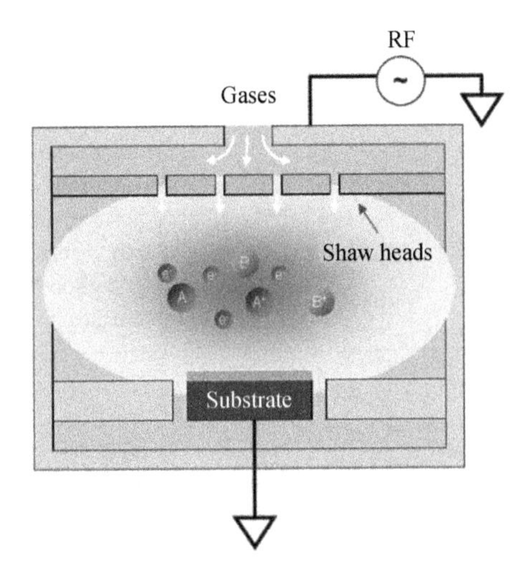

Figure 5.3 Schematic picture of plasma enhanced chemical-vapor deposition equipment

High-energy electrons can cause molecular excitation, ionization, decomposition, and formation of high-chemical-activity groups (i.e. activated molecules, atoms, ions, etc.) in the reaction gas to occur at temperatures as low as room on the substrate. Thus, plasma discharge has enhanced the chemical reaction, so this technique is called plasma enhanced chemical vapor deposition (PECVD). Table 5.1 lists some typical examples of the use of the PECVD method to make films.

Table 5.1 Some typical applications of PECVD [13]

Application	Composition	Gas raw material	Merits
Insulation and passivation film	SiO_2 Si_3N_4 P-PSG BSG Al_2O_3	SiH_4+H_2O $SiCl_4+O_2$ $Si(OC_2H_5)_4$ SiH_4+NH_3 $SiH_4+PH_3+O_2$ $SiH_4+BH_6+O_2$ $2AlCl_3+3CO_2+3H_2$	The temperature is low, the pinholes can be avoided, and impurities such as Na^+ can be avoided PSG film can be produced at lower temperatures
Amorphous silicon solar electron, Electrophotographic copier	α-Si	$SiH_4(SiH_2Cl_2)+B_2H_6$ (PH_3) or mixed gases: $SiH_4+SiF_4+Si_2H_6$	The substrate material does not need to be a single crystal, and the P-N junction can be conveniently prepared by changing the gas of the doped medium lower temperatures (20 ℃ ~ 400 ℃) Large-area film forming, low price.
Plasma polymerization	Organic compound		No need to destroy organic monomers completely. By choosing the conditions for the formation of clusters, polymers can be polymerized into organic compounds, and amorphous polymers cannot be obtained in the general way
Wear and corrosion resistant film	TiC TiN TiC_xN_{1-x}	$TiCl_4+CH_4$ $TiCl_4+N_2$ $TiCl_4+CH_4+N_2$	The film forming temperature is low, the film is even and smooth, the adhesion of film and substrate is strong, and the deposition rate is high
Other films	SiC Si, Ge Al_2O_3 GeO_2 B_2O_3 TiO_2 SnO_2 BN P_3N_6	$SiH_4+C_2H_2$ SiH_4+CH_4(or CF_4) SiH_4, GeH_4 $AlCl_3+O_2$ Alkyl or alkoxy compounds $B_2H_6+NH_3$ $P+N_2$	Film forming temperature is low; Film composition and properties can be adjusted; the film is even and smooth, with good surface quality

5.3 High-density plasma chemical-vapor deposition

Frequently used plasmas are DC, RF, and microwave plasmas. Note that the chemical-reaction process in PECVD is very complex, and the properties of the deposited films are closely related to the deposition conditions. Many parameters—such as operating frequency, power, pressure, substrate temperature, partial pressure of the reaction gas, reactor geometry, electrode space, electrode material, and pumping speed—may affect the film quality. Moreover, many factors interact with each other, and some are difficult to control. The reaction mechanism, reaction kinetics, and reaction process are not presently clear. Its application is also being developed and expanded.

In conventional plasma chemical-vapor deposition, DC or RF capacitive coupled discharge is used to form plasma. The plasma density is between $10^8 \sim 10^{10}$ cm^{-3} and the plasma density is low. In recent years, several high-density plasma-discharge techniques have been developed, including inductively coupled plasma and electron-cyclotron-resonance plasma, as shown in Figure 5.4. A common feature is that radio or microwave power is coupled to the plasma via a dielectric or dielectric wall. No power is coupled to the plasma by inserting electrodes into the reaction chamber, as in capacitive coupled-discharge plasma. This power-transmission mode lowers the plasma-sheath voltage to a typical value of 10~40 V. Thus, only a small amount of power dissipation in the sheath region accelerates the particles, and most of the power coupling yields electrons in the bulk plasma. Therefore, the energy can be efficiently used to ionize the reactant gas and increase the plasma density. On the other hand, the electrode placed on the substrate can also be driven by another RF power source, and the negative bias generated can control the ion energy of the bombarding substrate. Therefore, these high-density plasma systems can independently control the ion flux and the energy of the bombarding substrate, and these two parameters are of great importance for the deposition of thin films.

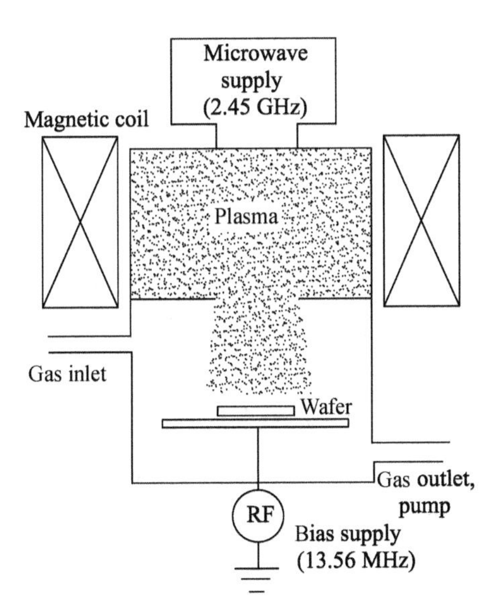

Figure 5.4 Schematic picture of high-density plasma chemical-vapor-deposition equipment

5.3.1 Inductively coupled plasma CVD

The RF power is fed into the non-resonant induction coil and the energy is delivered to the plasma by inductively coupled discharge. This offers the following advantages: high plasma density, simple structure, and no external magnetic field. Figure 5.5 shows two types of inductively coupled discharge-plasma devices.

The cylindrical device is shown in Figure 5.5(a). A coil is wound around the quartz cylindrical discharge and connected with an RF power supply. RF can also be applied to electrodes on the substrate to form a bias to control the energy of the bombarding ions.

The planar device is shown in Figure 5.5(b). The planar coil is spirally wound from the axis to the discharge chamber in the external direction and placed on the flat quartz dielectric window. To improve the uniformity of the radial plasma density, a multipolar permanent magnet can be set around the cavity to form a multi-level magnetic field to restrain the plasma. The coil can also be placed in the vacuum chamber near the base plate; this close-coupled plasma has a better uniformity, even without an additional permanent magnet. For large-discharge cells, the coils should be wound to improve the plasma's radial uniformity.

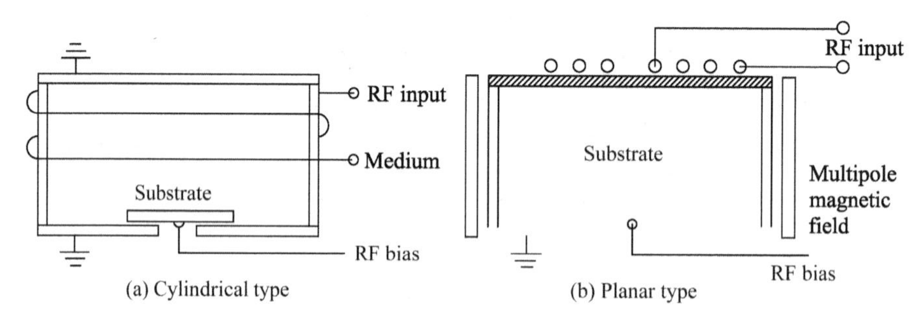

Figure 5.5 Two inductively coupled discharge systems

The frequency of the RF-power supply is generally 13.56 MHz and the output resistance is 50 Ω. It is connected to the induction coil to drive the discharge through the impedance-matching network. The coil can also be pushed by a balanced transformer. The earth terminal located in the middle of the coil can be connected to the transformer; this reduces the maximum voltage between the coil and the plasma, as well as the RF current generated by capacitive coupling. An electrostatic shield can be established between the coil and the plasma, which can further reduce the capacitance coupling impedance such that the induced electric field can be efficiently coupled to the plasma.

5.3.2 Electron-cyclotron resonance (ECR) plasma CVD

The microwave can produce a higher electric field in the resonant cavity, which can ionize the gas at low pressure to form a plasma. However, there is a limit to the plasma density, it produces: $n_p(m^{-3}) = 0.012 f^2$ (frequency f, Hz). To break the density limit, a magnetic field can be introduced. The introduction of the magnetic field B will cause the electrons to revolve. At this point, the input microwave frequency ω and the electron-cyclotron frequency $\omega_{ce} = eB/m_e$ produce what is known as electron-cyclotron resonance (ECR). Thus, the microwave energy will be highly coupled to the electron, resulting in the formation of high-density plasma.

Figure 5.6 is a sectional view of a typical ECR plasma CVD systems in which a microwave of 2.45 GHz is incident upon the discharge chamber via a quartz window in the direction of the field. An electromagnetic coil winding outside the discharge chamber produces a divergent magnetic field along the axis. When the magnetic-field strength reaches 875 Gs, the electron-cyclotron-resonance condition is satisfied, and the resonant coupling of the microwave energy is transmitted to form a high-density plasma. Since the magnetic field is divergent in the axial direction of the discharge chamber, as shown in Figure 5.6 (b), the plasma diffuses along the magnetic line of force to the deposition chamber with the substrate. Throughout the discharge, free radicals are generated with high densities, and these particles impinge upon the substrate to form a thin film.

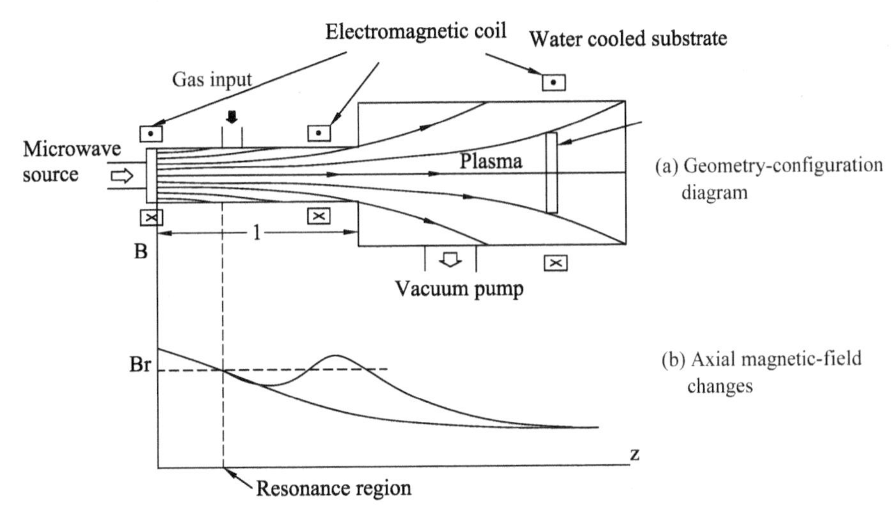

Figure 5.6 Typical ECR plasma CVD system[13]

Figure 5.7 presents a diagram of various ECR systems in different configurations. The systems in Figure 5.7 (a) have a high aspect ratio, and the plasma source is far from the substrate.

Microwaves travel in the direction of the magnetic field. The ECR region is annular or disk shaped, with a distance of about 50 cm from the substrate. When the plasma is transported from the resonance region to the substrate, the ion flux is reduced, and the collision energy between the ion and the substrate is increased. The device in Figure 5.7 (b) reduces the aspect ratio, and the ECR-resonance region enters the deposition chamber only 10~20 cm from the substrate. Owing to the strong relationship between the distribution of plasma uniformity and the axial magnetic field, a multipolar permanent magnet is placed in the deposition chamber to further improve the uniformity, as shown in Figure 5.7 (c). This increases the confinement of the plasma, thereby reducing losses. Another way to increase plasma density and uniformity is to connect the microwave source with the discharge chamber and to set the resonance region close to the substrate, as shown in Figure 5.7 (d). It can be seen that it has a low aspect ratio; the uniformity of the plasma can be improved by relatively flat and radially resonant regions.

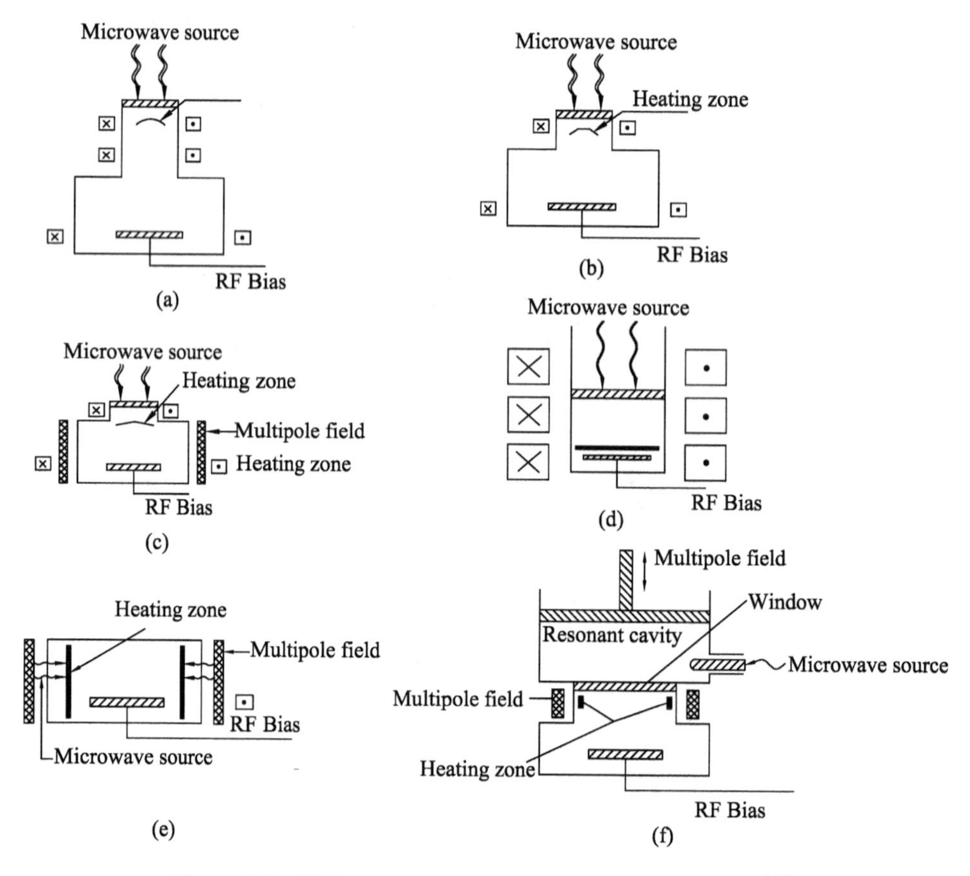

Figure 5.7 ECR systems with several different structures[13]

Figure 5.7(e) shows a DECR-discharge device with a multipolar distribution, and the microwave propagates vertically in the direction of a strong multipole magnetic field.

More than four microwave regulators are placed around the cavity to improve plasma uniformity; each microwave modulator produces resonance regions near the wall of the reactor. Figure 5.7(f) shows a microwave source with a tuner at the top and side to regulate the axial and radial distributions of the microwave, respectively. The gate located below the plasma-producing region is mainly designed to block microwave transmission and facilitate plasma diffusion. The linear-resonance region, similar to Figure 5.7(e), is generated by 8~12 permanent magnets distributed around the cavity.

5.4 Other chemical-vapor deposition

5.4.1 Metal-organic chemical-vapor deposition

Metal-organic CVD (MOCVD) is a technique for vapor-phase epitaxial growth of films by a thermal-decomposition reaction of organometallic compounds. It is mainly used for vapor-phase growth of compound semiconductor films. A schematic picture of the MOCVD system is shown in Figure 5.8.

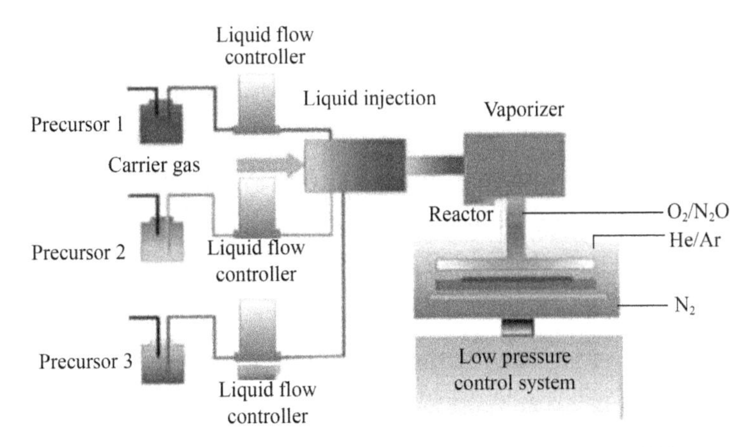

Figure 5.8 Schematic picture of the MOCVD system[15]

5.4.1.1 Principle of MOVCD

Compounds used as raw materials in the MOCVD method must meet the following conditions: (1) Under normal temperature conditions, they must be stable and easy to handle; (2) Reaction by-products should not hinder the crystal growth nor pollute the growth layer; (3) In order to realize vapor growth, they should have appropriate vapor pressure (> 100 Pa) around the room temperature.

Metal alkyl or aryl derivatives, alkyl derivatives, acetyl acetone compounds, and carbonyl compounds are often used as source materials. The leftmost elements of Ⅲ ~ V group and Ⅱ ~ Ⅳ group in the periodic elements table have strong gold property and cannot be an inorganic raw material compound that meets requirements; however, its organic compounds, especially alkyl compounds, mostly meet the requirements for raw materials. The elements on the right of the periodic elements table are strongly non-metallic, and their hydrides meet the requirements of the MOCVD compound. In addition, non-metallic alkyl compounds can also be used as MOCVD raw materials. Therefore, quite a few substances can be used as raw materials. For example, for GaAs, $Ga_{1-x}Al_xAs$, with Ga and Al raw materials, one can choose $(CH_3)_3Ga$ (three methyl gallium TMG), $(CH_3)_3Al$ (three methyl aluminum TMA); or $(C_2H_5)_2Zn$ (two ethane, zinc, DEZ). For As raw material, AsH_3 gas can be selected. Using the above material, a compound semiconductor can be obtained by thermal decomposition at high temperature, for example:

$$(CH_3)_3Ga+AsH_3 \xrightarrow{630\ ℃\sim 675\ ℃} GaAs_3CH_4;$$

$$(1-x)(CH_3)_3Ga+x(CH_3)_3Al+AsH_3 \xrightarrow{600\ ℃\sim 700\ ℃} Ga_{1-x}Al_xAs+3CH_4.$$

The MOCVD system can also be divided into low-pressure MOCVD and atmospheric pressure MOCVD; the work pressure of the former is generally $10^4 \sim 2\times10^4$ Pa.

5.4.1.2 Characteristics of MOCVD

MOCVD is a new epitaxial technology that has been developed rapidly over the past ten years. It has been successfully utilized in the fabrication of superlattice structures, ultra-high speed devices and quantum-well lasers. MOCVD is therefore determined by rapid development, mainly by its unique merits. MOCVD is characterized by the following features.

(1) Low deposition temperature. For example, the ZnSe film is deposited using ordinary CVD technology, the deposition temperature is about 850 ℃, and the MOCVD is only about 350 ℃. SiC is prepared from Tetramethylsilane with a growth temperature of less than 300 ℃. This is much lower than the growth temperature of low-pressure CVD (above 1,300 ℃) using $SiCl_4$ and C_3H_8 as its source. As the deposition temperature is low, self-pollution (boat material, substrate, reactor, etc.) is reduced, and the purity of the film is improved. Many wide-bandgap materials have volatile components that grow easily at high temperature and produce vacancies and form non-radiative transition centers. The presence of vacancies and impurities is responsible for self-compensation. Therefore, low deposition is beneficial for reducing vacancy density and solving the self-compensation problem, and the orientation requirement of the substrate is low.

(2) Because there is no halide material, there is no etching reaction in the process of

deposition. The deposition rate can be controlled by diluting the carrier gas, which is advantageous for depositing a large-thickness film, or many layers of polar film with different components (a few nanometers thick). Thus, various hetero-structures can be used to prepare super-lattice materials and epitaxial growth.

(3) The range of use is wide and almost all compounds and alloy semiconductors can be grown. For example, by controlling the ratio of the amounts of the organic metals of group III (Al, Ga, In and so on), mixed crystals of different compositions can be generated. But MOCVD is a non-balanced growth process; energy deposition of CVD and halogen liquid-phase epitaxy (LPE) mixed crystals cannot be obtained.

(4) The reaction device is easy to design and simpler than gas-phase epitaxy devices. The growth-temperature range is wide. In addition, the growth is easy to control and suitable for mass production.

(5) Epitaxial growth can be achieved on sapphire and spinel substrates.

The main drawbacks of MOCVD are:

(1) Although it is commonly used to replace the halide CVD by metal organic compounds and eliminates pollution and corrosion of the halogen, many toxic and flammable organic metal compounds are involved. Their preparation, storage, transportation and use are difficult. Therefore, strict protective measures must be taken.

(2) Because of the low reaction temperature, some organometallic compounds react in the gas phase. Solid particles are deposited on the substrate surface to form impurity particles in the film, destroying its integrity.

5.4.2 Light chemical-vapor deposition

Light chemical-vapor deposition is a chemical-vapor-deposition technique that uses light waves to break down gases, increase the chemical activity of reactive gases, and promote chemical reactions between them. In the process of film deposition, a certain activation energy is required to make the reactants undergo chemical reactions. When that activation energy is provided by light energy, only the radiation absorbed by the reactants can lead to photochemical reactions. For example, a variable-wavelength CO_2 laser can break down the vibrations of a SiH_4 or Ar^+ and decompose it. An energy of several electron volts is used to excite electrons directly and ionize molecules.

In addition to the thermal decomposition of the laser, the combined effects of excitation and heating can be generated simultaneously if the absorption spectrum of the reacting gas overlaps with the heated electromagnetic spectrum. By selecting the laser wavelength, several atoms in the polyatomic molecule can be interrupted to crack some particular chemical bond. The thin film is realized by chemical reaction.

The light sources of CVD include an Hg lamp, a transverse-excited atmospheric-pressure (TEA) laser, a CO_2 laser, a short-wavelength laser and ultraviolet light source, etc. UV light with a wavelength of 250 nm is adequate to break the chains of certain organometallic compounds and obtain metallic films. For example, $Al(CH_3)_3$ and He or $Cd(CH_3)_2$ and He mixed gas can be irradiated with UV to produce light decomposition at room temperature. A CO_2 laser with a wavelength of 10.6 m is particularly effective for heating substrates. The rate of Ni film prepared by $Ni(CO_4)$ is as large as 1~17 m/s, with a CO_2 laser beam of 0.5 to 5 W. Similarly, Fe, W, Al, Sn, Si, TiO_2, TiC, SnO_2, and other films can be obtained.

In addition, the use of light energy instead of heat energy in MOCVD can solve the problem of the deposition temperature being too high. This is called light MOCVD.

5.4.3 Atomic layer deposition

Atomic layer deposition (ALD), which is a chemical gas phase thin film deposition method based on alternate saturative surface reaction, as shown in Figure 5.9. As distinct from the other CVD techniques, in ALD the source, vapors are pulsed into the reactor alternately and separated by purging or evacuation periods. Each precursor exposure step saturates the surface with a monomolecular layer of that precursor. This results in a unique self-limiting film-growth mechanism with a number of advantageous features, such as excellent conformality/uniformity and simple/accurate film-thickness control. The current interest in ALD is largely centered on non-epitaxial films. ALD is a unique process based on alternate surface reactions, which are accomplished by dosing the gaseous precursors on the substrate alternately. Under ideal conditions, these reactions are saturative, ensuring many advantageous features, such as excellent conformity, large-area uniformity, accurate and simple film-thickness control, repeatability, and large-batch processing capability.

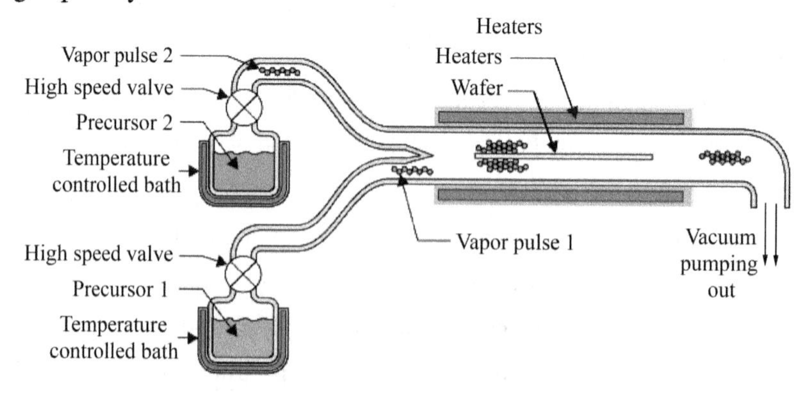

Figure 5.9 Schematic picture of the atomic layer deposition deposition system

Exercises

1. What is the working mechanism of CVD? Please draw pictures to explain.
2. What is the working mechanism of MOCVD? Please draw pictures to explain.
3. What is the working mechanism of ALD? Please draw pictures to explain.
4. What reactant gases might you use for making the following films by CVD: (1) ZrC; (2) TaN; (3) TiB_2?
5. Consider the data given below:

Reaction	A	B	C
$TiCl_4 = Ti + 2Cl_2$	180,700	1.8	−34.65
$2TiN = 2Ti + N_2$	161,700	—	−45.54
$SiCl_4 = Si + 2Cl_2$	155,600	3.64	−43.90
$SiC = Si + C$	14,000	1.3	−5.68
$C + 2H_2 = CH_4$	−16,500	12.25	−15.62
$1/2H_2 + 1/2Cl_2 = HCl$	−21,770	0.99	−5.22
$SiO_2 = Si + O_2$	215,600	—	−41.50

The values of A, B, and C are given for $G^0 = A + BT \log T + CT$ (G^0 in cal.) From these data, determine whether it would be thermodynamically feasible to form the following ceramic films by CVD at a temperature of 850 ℃. (a) TiN from the nitridation of $TiCl_4$; (b) SiC from the reaction between $SiCl_4$ and methane; (c) SiO_2 from the oxidation of $SiCl_4$.

6. Using the library or other sources, give examples of oxide deposition using each of the forms of CVD listed in this chapter.

Chapter

6

Kinetics of the Process of Thin Film Growth

The typical thin film growth process in the vacuum environment includes many kinetic physical and chemical reaction processes. These kinetics of thin film growth is important because it determines the quality of their surface. The knowledge of kinetics allows optimization of the epitaxial growth technique.

6.1 Four steps of thin film growth

These kinetic growth of thin film usually includes four steps: adsorption, surface diffusion, nucleation and structure development, as shown in Figure 6.1.

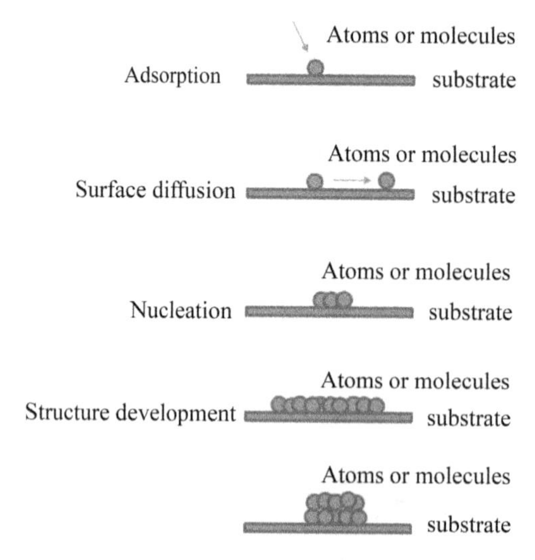

Figure 6.1 Schematic picture of four steps of thin film growth

In the following process, we will introduce these processes one by one.

6.2 Adsorption

6.2.1 Chemisorption and physisorption

Adsorption: Atoms and molecules arrive at substrate surface from the vapor phase.

A qualitative distinction can be made between chemisorption and physisorption in terms of their relative binding strengths and mechanisms, as shown in Figure 6.2. In chemisorption, a strong "chemical bond" is formed between the adsorbate atom or molecule and the substrate. In this case, the adsorption energy, E_a, of the adatom is likely to be comparable to or even greater than the sublimation energy of the substrate (typically a few eV/atom). Chemisorption involves sharing electrons in new molecular orbitals and is thus a form of strong bonding.

Physisorption is weaker, and is often considered to have no chemical interaction involved. The attractive interaction, in this case, is largely due to the van der Waals force. This force is due to fluctuating dipole (and higher-order) moments on the interacting adsorbate and substrate, and is present between closed-shell systems. Physisorption energies are on the order of 100 meV/atom. This is a weakly-adsorbed state in which a molecule impinges onto a surface.

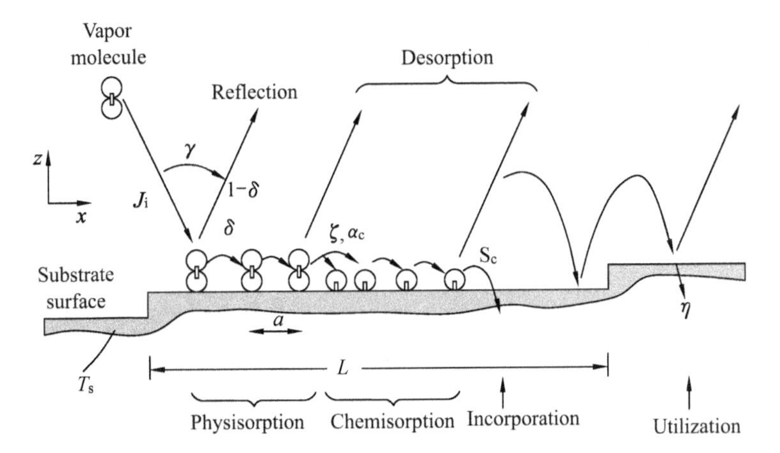

Figure 6.2 Schematic picture of chemisorption and physisorption during thin film growth[16]

6.2.2 Gas impingement on a surface

The impingement flux, Φ, is the frequency at which molecules collide with a surface, i.e. $\Phi = n\int_0^{\infty} v_x \mathrm{d}n_x$.

At thermal equilibrium, the Maxwell-Boltzmann velocity distribution is:

$$f(v_x)=\frac{\mathrm{d}n_x}{\mathrm{d}v_x}=\left(\frac{m}{2\pi kT}\right)^{\frac{1}{2}}\mathrm{e}^{-\frac{1}{2}mv_x^2/kT}.$$

So

$$\Phi = n\left(\frac{m}{2\pi kT}\right)^{\frac{1}{2}}\int_0^{\infty} v_x \mathrm{e}^{-\frac{1}{2}mv_x^2/kT}\mathrm{d}v_x = n\left(\frac{kT}{2\pi m}\right)^{\frac{1}{2}}. \tag{6-1}$$

Using the gas law, $p=nkT$, we have $\Phi=\frac{p}{\sqrt{2\pi mkT}}$. Substituting appropriate constants yields $\Phi=3.51\times10^{22}\frac{p}{\sqrt{MT}}$, where P is the pressure in Torr and M is the atomic number of the gas molecules.

How long does it take for a surface to be covered by a molecular monolayer? This time τ_c is the inverse of the impingement flux. If a monolayer consists of 10^{15} molecules,

$$\tau_c=\frac{10^{15}}{3.51\times10^{22}}\frac{\sqrt{MT}}{p}=2.85\times10^{-8}\frac{\sqrt{MT}}{p}. \tag{6-2}$$

In air at atmospheric pressure, $\tau_c=3.99\times10^{-9}$sec. In a vacuum of 10^{-10} Torr, $\tau_c=7.3$ hours.

6.2.3 Condensation

The adsorption of atoms/molecules is often preceded by condensation as shown in Figure 6.3. The adsorbates need to stay on the substrate surface long enough for the chemical reaction to occur (chemisorption) or simply stick there (physisorption).

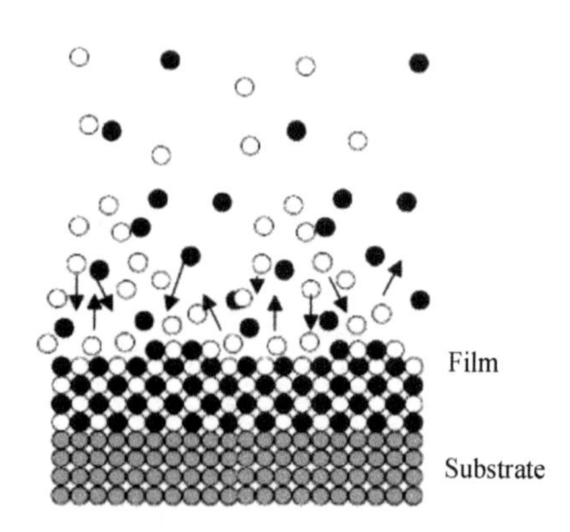

Figure 6.3 Schematic picture of the condensation process

At a certain temperature, there is an equilibrium vapor pressure (sublimation pressure); no deposition would occur at all unless one has super-saturation, which is normally achieved just above the substrate surface. Under super-saturation conditions, the density is so high, i.e. the atomic separation is so short. The condensation occurs due to van-der-Waals force as shown in Figure 6.4.

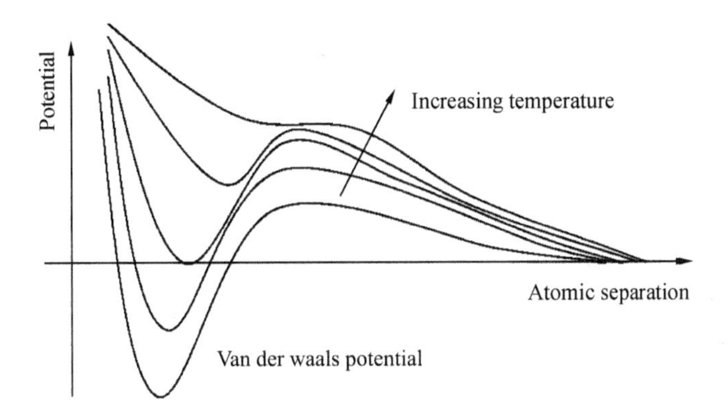

Figure 6.4　Schematic picture of the temperature-dependent van-der-Waals force between adatoms

6.3　Surface diffusion

Under a super-saturation situation, the atoms/molecules are condensed onto the substrate surface. The deposition rate or flux (of adatoms) is related to the pressure as $R=\frac{p}{\sqrt{2\pi mkT}}$. After the atoms are absorbed on the surface, they become adatoms with an (positive) adsorption energy E_a, relative to zero in the vapor. This is sometimes called the desorption energy. The desorption rate of the adatom is roughly given by $v_a e^{-(E_a/kT)}$, where v_a is the characteristic atomic-vibration frequency and is expected to vary relatively slowly (not exponentially) with T. Note that the desorption energy is assumed to come from lattice vibration only.

Adatoms can diffuse over the surface with energy E_d (migration-barrier energy) and corresponding frequency v_a (order of $10^{14}\ s^{-1}$). Since $E_d \ll E_a$, surface diffusion is far more likely than desorption. The probability that, during one second, the adatom will have enough thermal energy to surpass the barrier is $v_d e^{-(E_d/kT)}$. In a unit of time the adatom makes v_d attempts to pass the barrier, with a probability $v_d e^{-(E_d/kT)}$ of surmounting the barrier on each try. The adatom-diffusion coefficient (jumping a distance l, Figure 6.5) is then approximately $D=v_d l^2 e^{-(E_d/kT)}$.

This is, in fact, the mean-square displacement of the random walker per unit time, or the tracer diffusion coefficient. It is convenient to express surface areas in terms of substrate unit cells. Then, D becomes the number of unit cells visited by the adatom per unit time. The adatom lifetime before desorption is $\tau_a = v_a^{-1} e^{(E_a/kT)}$; then, the characteristic length within which the adatom can move is $L=\sqrt{D\tau_a}$.

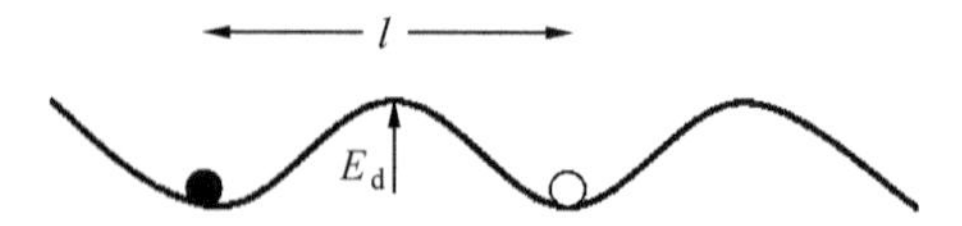

Figure 6.5 Schematic picture of the adatom's diffusion-migration barrier energy E_d on the substrate surface

6.4 Nucleation and growth of 2D islands

In the course of deposition, the atoms (monomer) arrive from the gas phase with a rate F (with units of atoms per surface unit cell per second, ML s^{-1}). ML stands for the monolayer. For simplicity, we assume the surface temperature is sufficiently low that only monomers diffuse on the surface and dimers remain immobile. As deposition proceeds, the number of dimers will increase roughly linearly until their concentration n_2 becomes comparable to the density of monomers, n_1. Thence, the probabilities for a diffusing monomer encounters one of its own or a dimer become comparable and cluster growth competes with the creation of new stable nuclei (dimers in our model). After the density of stable nuclei n_x(where x stands for any size that is stable) has increased sufficiently, any further deposition will exclusively lead to island growth. At this saturation-island density, the mean free path of diffusing adatoms is equal to the mean island separation, and the adatoms will have a much higher probability of attaching themselves to existing islands then of creating new ones. Approaching coverage of about half a monolayer, islands eventually coalesce, decreasing their density. A schematic picture of low-coverage and high-coverage of adatoms on the substrate surface is shown in Figure 6.6.

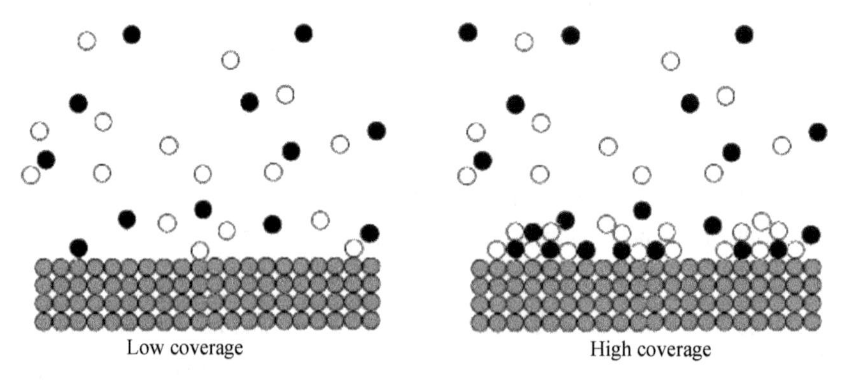

Figure 6.6 Schematic picture of low-coverage and high-coverage of adatoms on the substrate surface

If dimers are immobile and no re-evaporation occurs, then the rate equations for the densities of monomers and stable islands are

$$\frac{dn_1}{dt}=F-2\sigma_1 Dn_1^{\ 2}-\sigma_x Dn_1 n_x-\kappa_x F(Ft-n_1)-2\kappa_1 Fn_1,$$
$$\frac{dn_x}{dt}=\sigma_1 Dn_1^{\ 2}+\kappa_1 Fn_1. \tag{6-3}$$

The terms on the right-hand side of the first equation denote, respectively, the increase of monomer density due to deposition with flux F, its decrease due to the encounter of two diffusing adatoms resulting in the creation of a dimer (associated with the disappearance of two atoms), the decrease occurring when a monomer is captured by a stable island, and finally two terms denoting the decrease caused by direct impingement onto the stable island density n_x due to the creation of dimers, first when two monomers meet by diffusion, and then upon direct deposition onto an adatom. In these equations, coalescence is neglected; incorporation would add a further term, $-2n_x\left(F-\frac{dn_1}{dt}\right)$, to the second equation. In general, the problem is treated in the mean-field assumption—that is, outside the islands, the monomer density immediately takes on an average value. The time evolution of island and monomer densities can be obtained by integrating these equations. Very often, one is interested in the saturation-island density, as this reflects the mean free path for monomer diffusion. The temperature dependence of this quantity thus allows one to extract information concerning surface diffusion. The power law expressed by these equations leads to the approximation

$$\frac{D}{F}\approx\frac{L^6}{\ln(L^2)}. \tag{6-4}$$

The characteristic length L can either be identified with the mean island distance or with the mean free path of the diffusing adatoms before they create a new nucleus or are captured by existing islands. The logarithmic-correction term appearing in the denominator is small. Omitting this term yields $L\approx(D/F)^{1/6}$. Thus, L depends only on the ratio D/F. This is due to the fact that the flux is the only quantity introducing time if there is no re-evaporation and dimers are stable. Thus, the ratio of deposition to diffusion rate determines the mean free diffusion path and the mean island distance attained at saturation. The experimental examples of nucleation process on substrate surface are shown in Figure 6.7 and Figure 6.8.

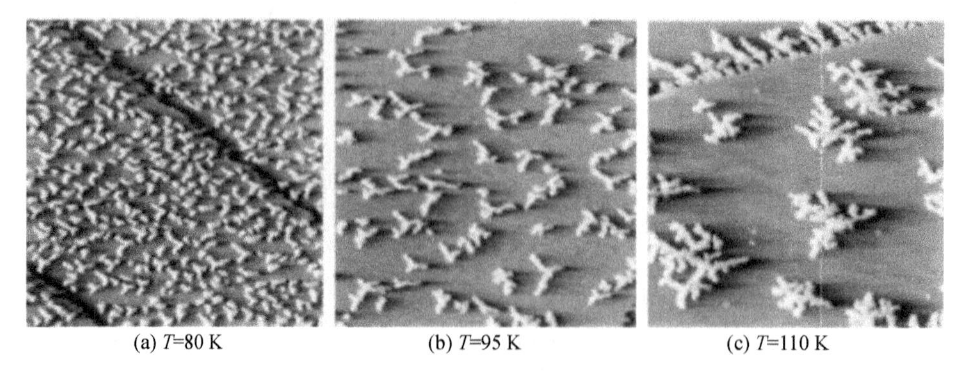

Figure 6. 7 STM images showing variation of the saturation-island density with temperature for deposition of 0.12 ML Ag onto a Pt(111) surface at 80, 95, and 110 K (F=0.001 1 ML/s)[18]

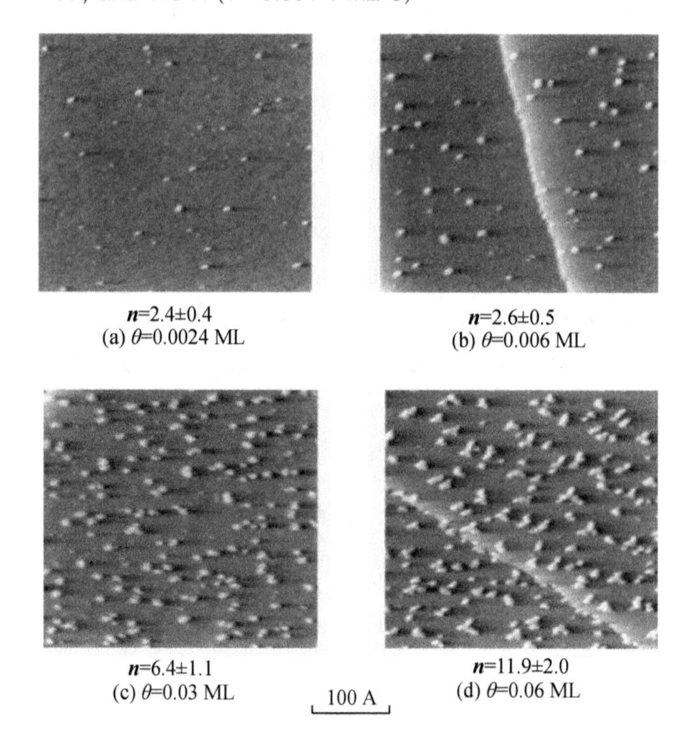

Figure 6.8 STM images showing the transition from the very early nucleation phase to island growth (2D) for Ag deposited (F=0.0011 ML/s) onto Pt (111) at 75 K[18]

If deposition continues, the nucleation centers become more numerous and the coalescence of 2D islands occurs. As the size of the 2D island is increased, the coverage of the substrate surface is enlarged. Ultimately, the concept of a film becomes imminent and we may now consider film growth, as shown in Figure 6.9.

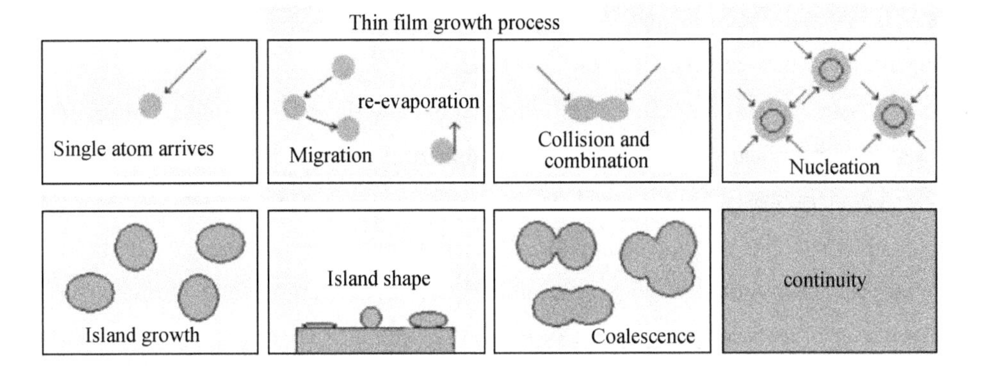

Figure 6.9 Schematic picture of the whole thin-film-growth process

6.5 Crystallization and film growth

6.5.1 Epitaxy

Epitaxy refers to the film-growth phenomenon whereby a relationship exists between the structure of the film and the substrate. In particular, it commonly denotes a single-crystalline layer grown on a single crystal surface. As shown in Figure 6.10, if the single-crystalline film and the single-crystalline substrate are of the same material, we call the growth homoepitaxy; if the film and the substrate are of different materials, we call the growth heteroepitaxy. For a defined relation between nucleus and substrate orientation, the nucleus must consist of at least 3 atoms for a 3-fold symmetry and 4 atoms for a 4-fold symmetry.

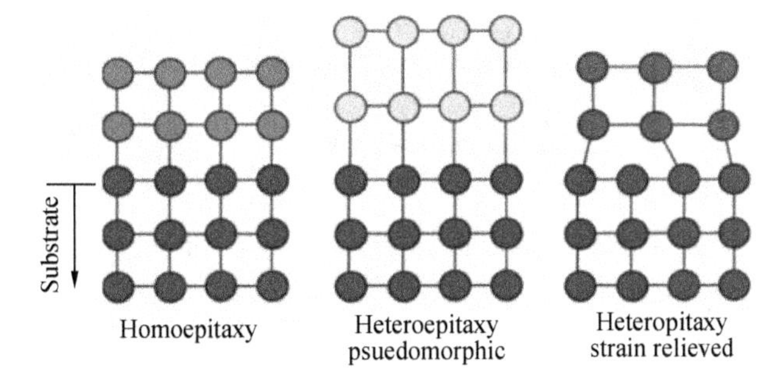

Figure 6.10 Schematic pictures of homoepitaxy and heteroepitaxy

6.5.2 Factors governing epitaxy

Three factors govern the thin film-epitaxy-growth process:

(1) Substrate. The substrate's structural compatibility can affect the thin film-epitaxy

process. The crystal structure and lattice constant of the substrate will lead to a lattice-matching problem between the film and substrate.

(2) Chemical compatibility. The chemical bonding and chemical diffusion at substrate/film's interface will finally affect the thin film-epitaxy-growth process.

(3) Temperature. Above a well-defined elevated-substrate temperature Te, good epitaxy is obtained. Te depends upon the deposition rate, particle energy, and surface contamination. The reason for the general need for a higher temperature is obviously the reduction surface contamination by desorption, the enhancement of the atomic surface mobility to reach favorable sites, and the enhancement of diffusivity in the deposition, thus favoring re-crystallization and defect annihilation.

6.6 Growth modes

The classification of three growth modes was first introduced by Ernst Bauer in 1958, as shown in Figure 6.11. The layer-by-layer (or Frank-van der Merwe) growth mode arises because the atoms of the deposited material are more strongly attracted to the substrate than they are to themselves. In the opposite case, where the deposited atoms are more strongly bound to each other than they are to the substrate, the island mode (or the Volmer-Weber mode) results. As an intermediate case, the layer-plus-island (or Stranski-Krastanov growth mode) is much more common than one might think. In this case, layers' growth starts first, after which the island mode may become dominant.

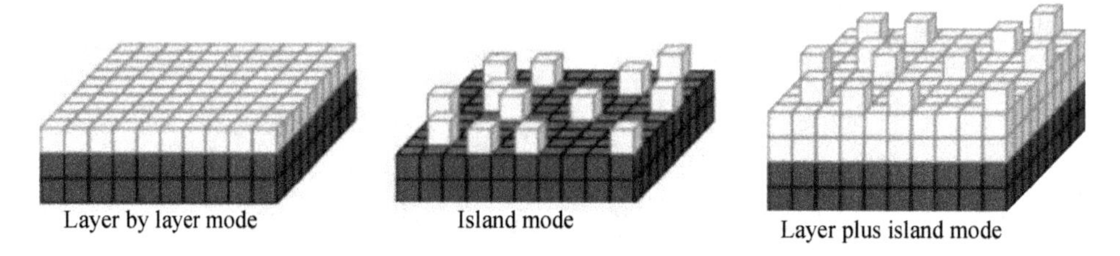

Figure 6.11 Classification of three growth modes of deposited films

Lattice mismatch has a marked effect upon film morphology. The strain resulting from lattice mismatch contributes to the interface energy, a key parameter determining the growth mode. However, the surface free energies for the substrate and film materials also influence the mode of growth. For heteroepitaxy in general, observed growth modes have been placed in three categories depending on the resulting film morphology: (1) the Frank-van der Merwe (FM) or layer-by-layer mode, (2) the Volmer-Weber (VW) or 3D island mode, and (3) the Stranski-Krastanow (SK) or 3D island-on-wetting-layer mode. The interatomic interactions between substrate and film materials are stronger and more attractive than those between the different atomic species within the film material in FM growth, whereas just the opposite is true in VW growth. SK growth occurs for interaction strengths somewhere in the middle.

Bauer and van-der-Merwe have cast the energetics of film growth into a particularly simple form under the assumption of equilibrium between the film-energy components in the gas phase and those on the film surface. The thin film nucleus will adjust itself to minimize the total surface energy:

$$\sum_k \gamma_k A_k = \text{Minimun}, \tag{6-4}$$

where γ_f and γ_s are the respective surface free energies of the film and substrate and γ_i is the interfacial free energy between the film and substrate, as shown in Figure 6.12.

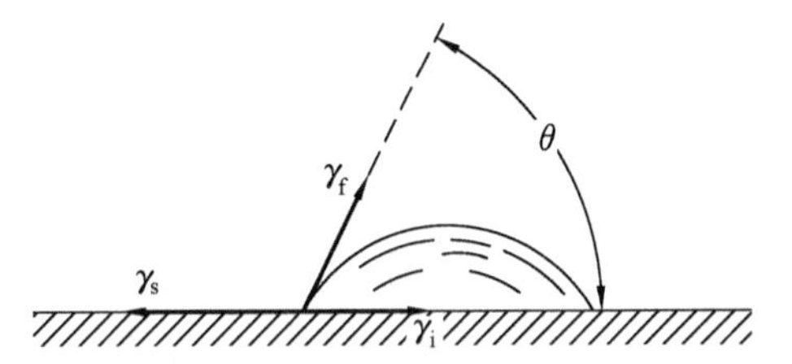

Figure 6.12 Classification of the three growth modes of deposited films[16]

In this formalism, the layer-by-layer growth of the film on the substrate requires that

$$\Delta\gamma = \gamma_f + \gamma_i - \gamma_s \leqslant 0. \tag{6-5}$$

The latter quantity depends upon the strain and the strength of chemical interactions between film and the substrate at the interface. The above equation says that the sum of the film's surface energy and the interface energy must be less than the surface energy of the substrate for wetting to occur. Alternatively, it becomes easier for layer-by-layer growth to occur as the substrate's surface energy increases. Thus, FM growth is expected if the above equation is obeyed. However, the strain energy, which is a term in γ_i, increases linearly with the number of strained layers. At some thickness, $\gamma_f + \gamma_i$ exceeds γ_s and the growth mode transforms from layer-by-layer to 3D island-on-wetting-layer, resulting in 3D islands on the 2D wetting layer. Alternatively, γ_f may be sufficiently in excess of γ_s that the above equation is never satisfied, even for a strong attractive interaction between the film and the substrate with little strain ($\gamma_i < 0$). In this case, 3D islands nucleate from the onset, resulting in V-island growth. Finally, in the limit of zero lattice mismatch and weak chemical interactions between A and B at the interface ($\gamma_i \cong 0$), the growth mode is determined entirely by the surface free energies of the film and substrate material. An interesting consequence of the above equation is that growing super-lattices (A/B/A/

B ...) consisting of perfectly laminar films and atomically flat interfaces are unlikely unless $\gamma \cong 0$. If $\Delta\gamma \leqslant 0$ for A on B, then $\Delta\gamma > 0$ for B on A. However, the clever use of surfactants, which have the effect of lowering the surface energy of the high-surface-energy component, can alleviate this problem.

6.7 Structure development

The surface nucleus grows larger to form a continuous film with grain and grain boundaries. Firstly, the coalescence of surface nuclei forms a continuous film, and the nucleation step of film growth is completed. There are many factors affecting the growth of thin films, among which the main factor is thermal effect (substrate temperature).

We can define a ratio factor T_s/T_m (in K), where T_s is substrate temperature and T_m is melting point of film. Three structural zones (Z1, Z2, Z3) can be identified according to the shape of grain growth, as shown in Figure 6.13 and Figure 6.14. Besides, ZT is a transition between Z1 and Z2. The data presented in Figure 6.14 and Figure 6.15 are obtained from thermal/electron beam evaporation techniques. But it can be applied for other evaporation techniques, such as sputtering deposition. Moreover, it is applicable to both metals and ceramics. We will analyze the above four structures zones (Z1, ZT, Z2, Z3) one by one.

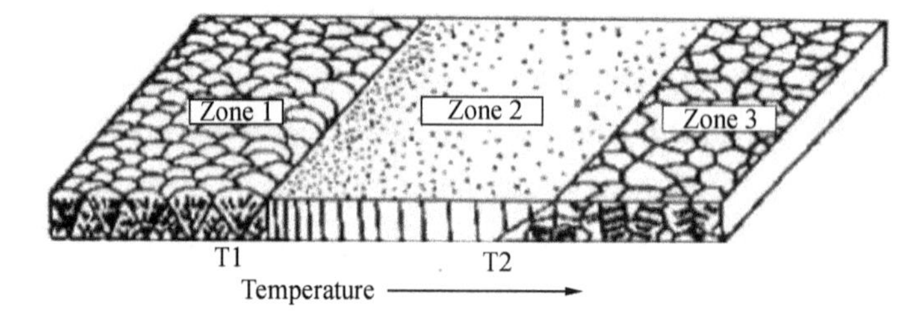

Figure 6.13 Influence of substrate temperature upon the thin film growth process[16]

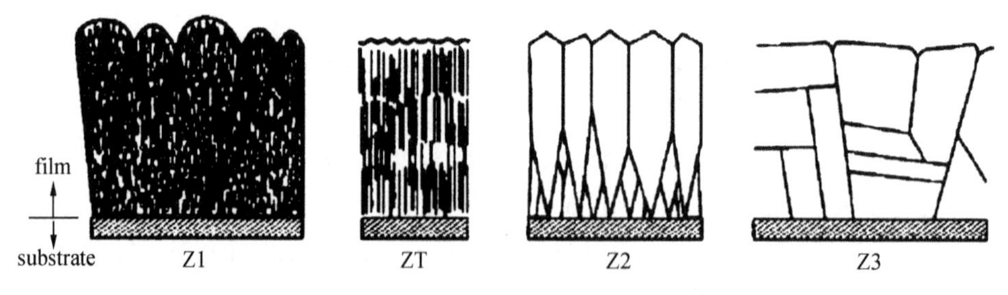

Figure 6.14 Characteristics of four basic structural zones in cross-section view[16]

Z1 occurs at low T_s/T_m. It has the following characteristics:

(1) Surface diffusion negligible;

(2) Poor crystallinity in columns (possibly amorphous);

(3) Columns separated by voids;

(4) Thicker films will superimpose onto structure, forming array of cones;

(5) Cones terminate in domes;

(6) Size of domes increase with film thickness.

ZT is similar to Z1 but associated with energy-enhanced process (i.e. plasma, etc.). Its surface diffusion can be ignored. It has no voids and domes.

Z2 occurs at T_s/T_m>0.3. It has the following characteristics:

(1) Surface diffusion is significant;

(2) Columns have tight grain boundaries—diameter of columns increases with T_s/T_m;

(3) Crystalline columns has less defect than with Z1 or ZT;

(4) Facets observed at surface of film;

(5) It can occur in amorphous films.

Z3 occurs at T_s/T_m>0.5. It has the following characteristics:

(1) Considerable Bulk annealing present during film growth;

(2) More isotropic/equiaxed crystallite shapes;

(3) Film surfaces are smoother;

(4) Grain boundaries can develop grooves.

It must be mentioned that all the above four zones cannot always be identified (especially Z3). Firstly, transition from one zone to another is not abrupt with temperature. Secondly, epitaxial films do not exhibit these film structures. Finally, surface topography will vary according to other deposition conditions. For example, for sputtering deposition, the transition zones also depend on the Ar pressure used, as shown in Figure 6.15.

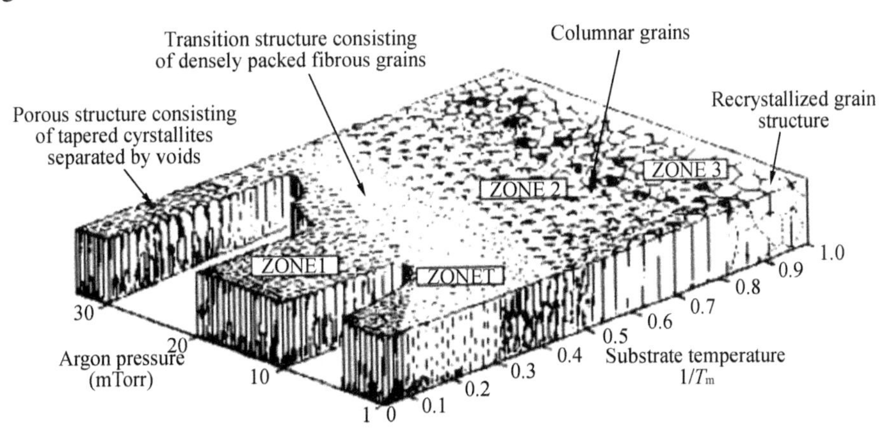

Figure 6. 15 Influence of substrate temperature and argon pressure upon the microstructures of sputtering-deposited metal thin films[16]

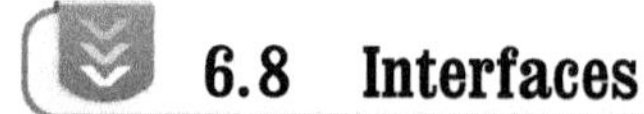

6.8 Interfaces

When two different films of different structure and elemental composition are deposited one on top of another (or single film with the substrate), an interface is inevitably involved, as shown in Figure 6.16. However, most materials "dissolve" in each other, including solid state, thus forming compounds.

Generally, it is hard to predict the extent of films interaction at their interfaces. The abrupt interface in Figure 6.16(a) is usually observed in Z1 and ZT phases when surface diffusion negligible. Usually, abrupt interfaces is "most" desirable in film deposition.

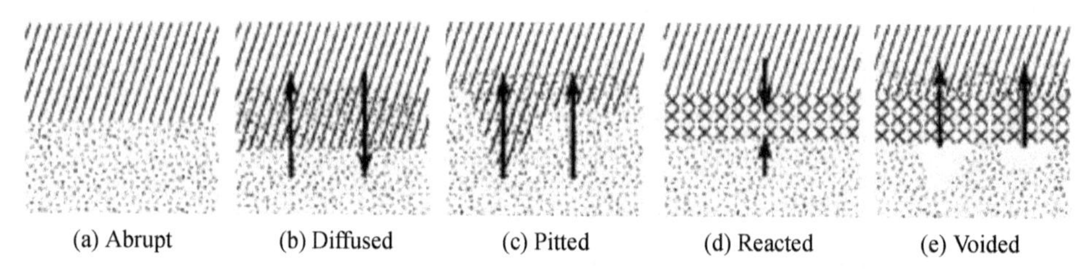

Figure 6. 16 Schematic pictures of different types of interface diffusion (arrows indicate direction of diffusion) [16]

But the inter-diffusion rate at interfaces vary depending on the film structure and the rates of solid-state diffusion. The phase diagram of film materials is able to give some help to identify. For example: Al film was deposited on Si substrate. The Al/Si interface will depend on the solubility of Al and Si, as well as substrate temperatures.

From phase diagram in Figure 6.17, both elements have limited solubility in each other. There is maximum solubility at 1.5% Al and 577 ℃. So Si can diffuse into Al at this temperature, forming interface pits [Figure 6.16 (c) and Figure 6.18] in Si substrate due to Si diffusing into Al film.

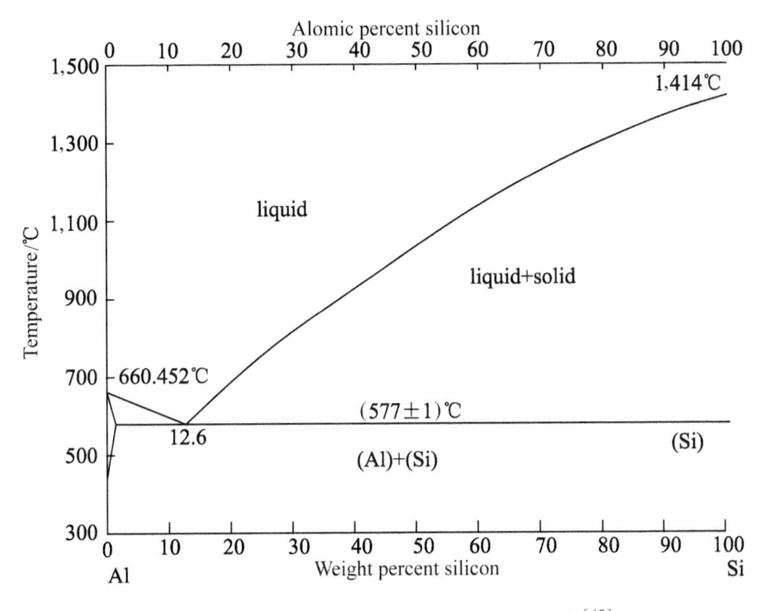

Figure 6.17　Phase diagram of Al and Si[16]

Figure 6.18　SEM image of the interface pits in Si(001) substrate after removing of Al film on top of the Si substrate[16]

Exercises

1. What are the four key steps for thin film growth on a substrate?
2. Please draw a schematic picture of the whole thin film-growth process.
3. What's adsorption? What's the difference between chemisorption and physisorption?
4. What's surface diffusion? What factors are the speed of surface diffusion depended on?

5. What's nucleation? What factors are the rate of nucleation depended on?
6. What are the three important growth modes of thin films? Please draw a picture to explain.
7. What's structure development? How are the structure zones defined?
8. Why are the interfaces of two different films so important? What types of interface diffusion can happen?

Chapter 7 Measurement of Film Thickness and Deposition Rate

Film thickness is an important parameter for characterizing the film, because many characteristics of the film are related to this quantity. In many applications, precision of film thickness is very high; for example, the control accuracy of integrated-circuit devices is often at nanometer level. Although thickness is a very conventional physical quantity, the problem of film thickness is not simple.

Thickness is a geometric concept, which refers to the distance between two parallel planes. The thickness of the film refers to the distance between the substrate surface and the film surface. The problem is that the actual surface of the film is not an abstract geometric plane, but is rather made up of atoms (or molecules) arranged in a certain way; its surface is not completely smooth, but may have holes, impurities, lattice defects, or even discontinuities, especially in the early stage of film formation or in the case of ultra-thin films. Additionally, the surface of the film may also be covered by adsorbed molecules, making it difficult to determine the film's surface. There are different thicknesses at different positions of the film, so it is difficult to give the film's exact thickness.

In addition, various methods of measuring film thickness adopt different technologies or principles in practice, and the measurement accuracy also differs. Therefore, for the same film, different measurement methods may give different results, i. e. different thickness values. Therefore, when the measurement results of film thickness are given, the measurement method used should be indicated. The measurement methods of film thickness can be divided into optical, mechanical, and electrical methods. Some of these methods involve non-destructive testing, while others will cause some damage to the film, which is called destructive testing.

Most methods of measuring film thickness must take the film out of the reaction chamber for off-line measurement. In fact, the properties and structures of the films depend upon the deposition conditions, such as the deposition method, temperature, and rate. After choosing the deposition method and temperature, it is very important to control the deposition rate on line to control the film's thickness. Therefore, whether the film thickness is uniform, whether it is consistent with the preset value, whether the thickness deviation is within the specified range, are all of key importance in determining the film's characteristics. The online measurement and monitoring of film thickness is a basic detection projects in the film-manufacturing industry, and the corresponding measurement technology is becoming more mature. In addition, the film-thickness and deposition-rate parameters are also important to consider when analyzing film characteristics.

The following sections will introduce several commonly used film-thickness-measurement methods.

7.1 Optical method

Various optical methods are widely used in the measurement of film thickness because they can be used not only to measure transparent films, but also opaque ones. They are easy to use and can achieve a very high measurement accuracy. When the thickness of the film is given, the refractive index, thickness uniformity, and other parameters of the film are often obtained at the same time.

7.1.1 Method based on measuring the absorption coefficient of light intensity

When a thin film with a light-transmission-absorption coefficient of α and a thickness of d has an intensity of I_0, its light intensity I becomes:

$$I=I_0(1-R)^2\exp(-\alpha d). \tag{7-1}$$

Here, R is the reflectivity of light at the film-air interface. Obviously, by measuring the change of light intensity, the thickness of the film can be determined by the above formula. This method is very simple and often used in the evaporation of metal film. When the deposition rate is fixed, the relationship between the transmission-light intensity and time is linear on the semi-logarithmic coordinate graph, making this method suitable for the on-line control of the film-deposition process and detection of film-thickness uniformity. However, it must be noted that only materials capable of forming continuous microcrystalline film when the thickness is very thin (such as Ni-Fe alloy) have this characteristic. For example, when silver is very thin (~30 nm), the intensity of transmission light decreases linearly with increase of the thickness of the film. Only when the thickness of silver film is greater than 30 nm is there an exponential-decline relationship. Therefore, it is necessary to check whether Formula (7-1) is valid for each material, or to determine its calibration curve if necessary.

7.1.2 Optical interferometry

The interference of the film can be explained in Figure 7.1. It is assumed that a film with a thickness of d and a refractive index of n_2 is deposited on a substrate with a refractive index of n_3. The top of the film is air (refractive index $n_1 = 1$). From a monochromatic light source of wavelength λ, the light S (light 1) that is irradiated on the surface of the film at a point "A" will be partly reflected by the interface (producing light 2). The other part is refracted into the film (producing light "ab"). A beam of light refracted into the film will be reflected again at "b" on the interface between the film and the substrate (producing light "bc") and refraction (producing light 3). Obviously, light beams 2 and 3 are wavelets originating from incident light 1, and both satisfy the coherent condition; interference occurs after they meet at point P. P is the maximum or minimum interference, which depends on optical path Δ of the two rays. It is not difficult to calculate the optical-path Δ difference:

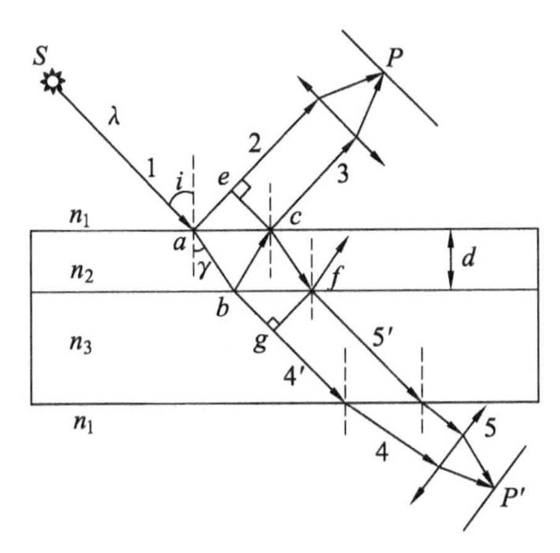

Figure 7.1 Thin-film-interference diagram

The interference phenomenon of the film can be explained by Figure 7.1. It is assumed that a thin film of thickness d and refractive index n_2 is plated on the substrate with a refractive index of n_3, and the upper part of the film is air (refractive index $n_1 =$ 1). Part of the light (light 1) from the monochromatic light source S with wavelength λ to a point a on the surface of the film will be reflected by the interface (producing light 2), and the other part will enter the film after refraction (producing light "ab"). The beam "ab" refracted into the film will reflect (produce light "bc") and refract (producing light 3) again at point b on the interface between the film and the substrate. Obviously, light 2 and light 3 are wavelets from incident light 1; they meet the coherence condition and they will interfere when they meet at point P. Whether the point P interference is a maximum or a minimum depends upon the optical-path difference Δ of the two rays. It is not difficult to calculate, and the optical-path difference Δ can be expressed as:

$$\Delta = n_2(\overline{ab}+\overline{bc}) - n_1\,\overline{ae}.$$

Considering

$$\overline{ab}=\overline{bc}=\frac{d}{\cos\gamma},\quad \overline{ae}=\overline{ac}\sin i = 2d\ \mathrm{tg}\,\gamma\,\sin i,\quad n_1\sin i = n_2 \sin\gamma.$$

We obtain

$$\Delta = 2n_2 d\cos\gamma = 2d\sqrt{n_2^{\ 2}-n_1^{\ 2}\sin^2 i}, \tag{7-2}$$

where i is the incident angle of the light, γ is the refraction angle in the film, and d is the thickness of the film near the incident point.

Generally, air or vacuum is considered above the film (refractive index $n_1=1$), and Equation (7-2) can be expressed as follows:

$$\Delta = 2n_2 d\cos\gamma = 2d\sqrt{n_2^{\ 2}-\sin^2 i}. \tag{7-3}$$

When $\Delta = k\lambda$, there will be interference maxima at point P, where k is the order of interference maxima, taking a positive integer; for each point corresponding to $\Delta = (2k+1)\dfrac{\lambda}{2}$, there will be interference minima.

When light is vertically incident and reflected, Equation (7-1) is simplified as follows:

$$\Delta = 2n_2 d. \tag{7-4}$$

When $\Delta = 2n_2 d = k\lambda$, point P will have an interference maximum of order k, taking a positive integer. When $\Delta = 2n_2 d = (2k+1)\dfrac{\lambda}{2}$, point P will have an interference minimum.

In practical applications, the problem of half-wave loss should be considered. Both theory and experiment have proved that, when light is incident from a light-sparse medium to a light-dense medium (that is, a medium with a large refractive index), it will produce a π-phase mutation equivalent to an additional optical path of $\lambda / 2$. For the medium series shown in Figure 7.1, we can prove that only when $n_1<n_2$, $n_2>n_3$ or $n_1>n_2$, $n_2<n_3$ is there half-wave loss between lights 2 and 3; the corresponding optical-path difference [Equation (7-2), (7-3), (7-4)] is added with $\lambda/2$ and the other analysis is unchanged.

The transmitted lights 4 and 5 also satisfy the coherence condition. Considering the problem of half-wave loss, the interference fringes of transmitted and reflected light are complementary. However, the spacing and shape of the stripes are the same. Whether the intensity of the interference fringes of the reflected or transmitted light is chosen to measure the fringes spacing depends on whether the reflected or transmitted light of the film is strong enough, and also on the transparency of the substrate used.

7.1.2.1 Foreign-economic-cooperation-office (FECO) method

If the substrate and the film are transparent but the reflected or transmitted light is sufficiently strong, the reflected and transmitted light intensities will change periodically with the thickness of the film in the simplest case when a monochromatic light of wavelength λ is vertically incident, showing a series of maxima and minima. The corresponding film thickness can be measured by recording the change of light intensity.

Figure 7.2 shows that the reflectivities of films with various refractive indices deposited on a glass substrate with a refractive index of 1.5 change periodically depending its thickness, and the medium above the film is air ($n_1 = 1$). Assuming that $n_1<n_2<n_3$, the reflectivity will decrease with the increase of film thickness after the deposition begins; when the optical thickness (n_2d) of the film reaches $\lambda/4$, the reflectivity becomes minimal. If the deposition continues, the reflectivity increases with film thickness, reaching a maximum value when the optical thickness of the film reaches $\lambda/2$; when the optical thickness of the film is $3\lambda/4$, the reflected light takes on a minimal value again; when the optical thickness of the film is λ, the reflected light is maximal. The opposite is true for transmitted light. If $n_1<n_2$, $n_2>n_3$ or $n_1>n_2$, $n_2<n_3$, due to the existence of half-wave loss, when the optical thickness of the film is $\lambda/4$, $3\lambda/4$, the reflected light will have a maximum value. When the optical thickness of the film is $\lambda/2$, λ and so on, the reflected light will have a minimal value. In the process of film deposition, the thickness of the film can be monitored by recording the number of times

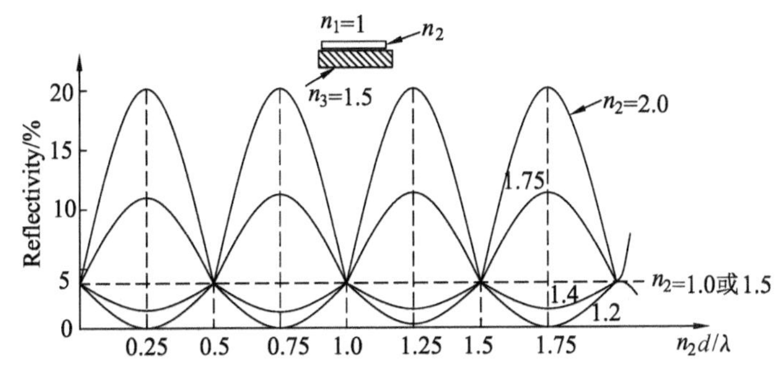

Figure 7.2 Reflectivity of the film and the change of optical thickness *nd*[13]

that the reflectivity of the deposited film passes through the extreme point. The deposition process should be interrupted when the reflectivity reaches a certain extreme value. For example, the optical thickness of the film is exactly equal to $N \cdot \lambda/4$ when the reflectivity passes through the extremal point N times.

If white light is used (in fact, a polychromatic light source is used), the reflected light intensity increases for those wavelengths whose optical thickness of the thin film is equal to odd multiples of the quarter wavelength, and decreases for even multiples. Therefore, the reflected color of the film is equivalent to the combination of these wavelengths with increased intensity. When the same film is observed by transmission light, its color is complementary to that of reflection. If we make observations under white light (polychromatic light source) when preparing the film (for example, by the evaporation method), the film will show different colors with the increase of its thickness: successively purple, blue, blue-green, green-yellow, yellow, orange, red, purple-red, etc. By observing the film's change in color, once the required color appears after a certain number of cycles, evaporation will stop and the required thickness of the film can be prepared.

Another method for using monochromatic light incident to meet the interference conditions by changing the incidence (and reflection) angles is called the variable-angle interference method (VAMFO). A schematic diagram of the measuring system is shown in Figure 7.3. In the process of continuous change of the sample angle, the alternation of interference maxima and minima can be observed under an optical microscope. Suppose $n_1<n_2<n_3$, $n_1=1$; from Equation (7-3), it can be concluded that when the reflection is great:

$$d=\frac{k\lambda}{2\sqrt{n_2^2-\sin^2 i}}. \tag{7-5}$$

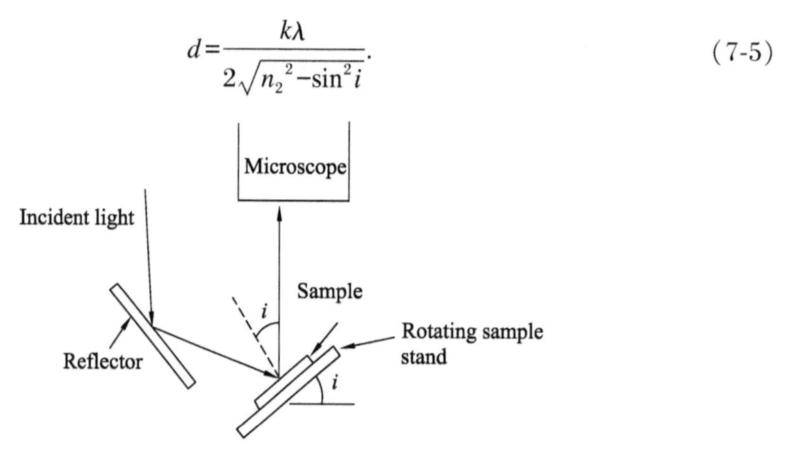

Figure 7.3 Principle of measuring film thickness by changing the angle[13]

Here, the interference series k can be calculated by observing the maximum number of times that the incident angle i starts from 0 (vertical incidence).

7.1.2.2 Equal-thickness interference-fringe (FET) method

If the film to be studied is opaque, a mask can be placed on the substrate in advance (the simplest available silicon-chip fragments), so that a step-shaped film will be produced during coating. Then, the mask is removed and a thin film (such as Ag or Al) with the highest reflectivity is completely and evenly plated. The reflectivity of the metal film is the decisive factor for producing distinct interference fringes and plays a key role in improving the measurement accuracy. That is to say, the height (thickness) of the film step can be easily measured using equal-thickness or equal-color interference fringes.

As shown in Figure 7.4 (a), the air gap (wedge) between two optical plates with small dip angles forms a thin layer for interference, and one of the optical plates is coated with a film to be measured. A highly reflective metal layer is uniformly deposited on the steps of the film. Considering the case where $n_1 < n_2 < n_3$ and monochromatic light is irradiated, Equation (7-4) shows that the maximum-interference condition for light is that the optical-path difference from the film (or substrate) is an integer multiple of the optical wavelength λ; that is:

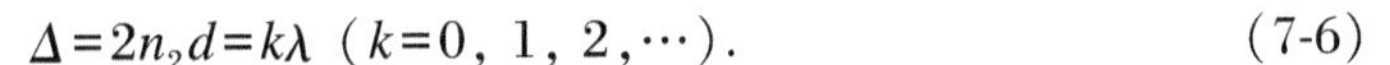

$$\Delta = 2n_2 d = k\lambda \ (k=0, 1, 2, \cdots). \tag{7-6}$$

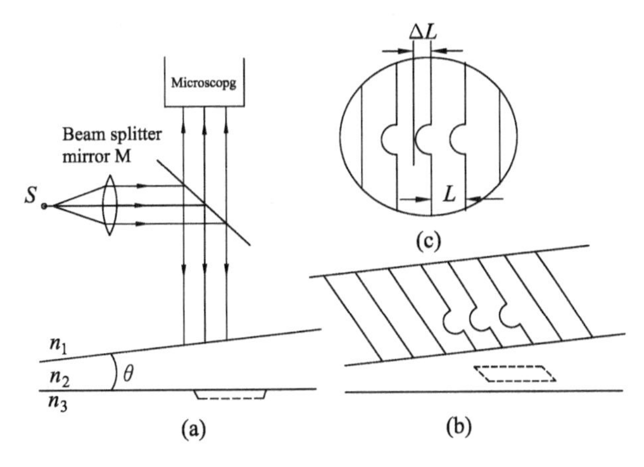

Figure 7.4 Measurement of film thickness by the equal-thickness interference-fringe method[13]

It can be seen that the optical-path difference at this time is completely determined by the thickness of the film at which the incident point is located, and the same thickness corresponds to the same level of interference fringes; that is, equal-thickness (interference) fringes are generated. The film-thickness difference between two adjacent interference fringes is as follows:

$$\Delta d = \frac{\lambda}{2n_2}. \tag{7-7}$$

A set of parallel straight stripes (equal thickness) with equal spacing will appear on the surface of the optical plate above, as shown in Figure 7.4 (b), and the stripe spacing is L. When the thickness of the step is increased, the stripe will bend towards the wedge direction. ΔL is the bending degree of the stripe, as shown in Figure 7.4 (c); then, the film thickness can be given by the following formula:

$$d=\frac{\Delta L}{L}\cdot\frac{\lambda}{2n_2}. \tag{7-8}$$

It should be noted that there is a common feature (defect) in all the optical methods for measuring the film thickness, that is, the optical thickness (nd) of the film is measured directly. To obtain the film thickness, the refractive index n_2 of the film must be known in advance. Sometimes, the refractive index of the same material is used. Otherwise, it is necessary to first assume a refractive index and then use its parameters such as the incident angle of a series of lights corresponding to the measured interference extremum, using the relevant software to continuously fit, and finally obtain the film thickness.

7.1.3 Ellipsometry

Ellipsometry's basic measuring principle is that the linearly polarized light produced by the polarizer becomes a special elliptically polarized light after being oriented to a certain 1/4-wave plate. When it is projected onto the surface of the sample to be measured, as long as the polarizer takes the appropriate light-transmission direction, the light reflected from the surface of the sample to be measured will be linearly polarized. According to the change of the polarization state (amplitude and phase) of the polarized light before and after reflection, we can use a more complex mathematical equation derived from electromagnetic theory to determine the film's thickness and refractive index, as well as optical parameters such as the material's absorption coefficient or the metal's complex refractive index. This principle was first put forward by Brewster, Fresnel and others and has a history going back more than 100 years. However, due to the difficulties in mathematical processing, it did not develop until the computer appeared in the 1940s. After decades of continuous improvement, ellipsometry has gone from manual to fully automatic, with a variable-incidence angle and variable wavelength especially suitable for real-time monitoring, which greatly promotes the development of nanotechnology. Ellipsometry has a high measurement accuracy (one to two orders of magnitude higher than general interferometry) and a high measurement sensitivity (it can detect the thickness change of a growing film of less than 0.1 nm). With the rapid development of computers and related software, optical fibers, and spectrometers, ellipsometry has been widely used in semiconductor materials, optics, chemistry, biology, and medicine.

Let the sample to be tested be a transparent isotropic film uniformly coated on the substrate. As shown in Figure 7.5, n_1, n_2, and n_3 are the refractive indices of the environmental medium, film and substrate, respectively, d is the thickness of the film, the incident angle of the incident beam on the film is φ_1, and the refraction in the film and the substrate have angles φ_2 and φ_3, respectively. According to Snell's law:

$$n_1 \sin \varphi_1 = n_2 \sin \varphi_2 = n_3 \sin \varphi_3. \quad (7\text{-}9)$$

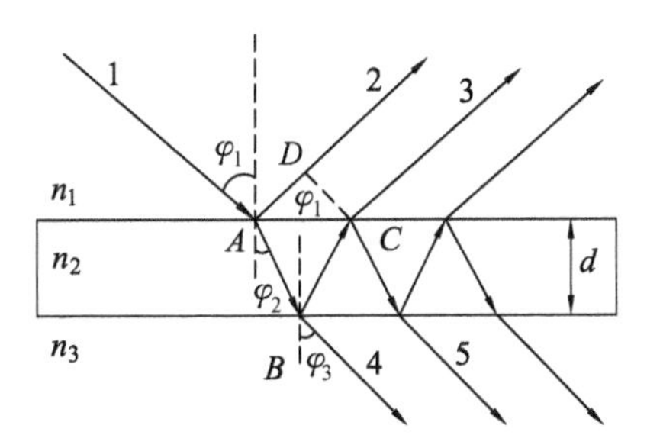

Figure 7.5 Reflection and refraction of an incident-light beam upon the surface of a sample to be measured

The electric vector of light is decomposed into two components, namely the p component of the vibrating surface in the plane of incidence and the s component perpendicular to this plane. According to the law of refraction and the Fresnel-reflection formula, the complex amplitude reflectances of the p and s components at the first interface can be obtained:

$$r_{1p} = \frac{n_2 \cos \varphi_1 - n_1 \cos \varphi_2}{n_2 \cos \varphi_1 + n_1 \cos \varphi_2} = \frac{\operatorname{tg}(\varphi_1 - \varphi_2)}{\operatorname{tg}(\varphi_1 + \varphi_2)}, \quad r_{1s} = \frac{n_2 \cos \varphi_1 - n_1 \cos \varphi_2}{n_2 \cos \varphi_1 + n_2 \cos \varphi_2} = \frac{\sin(\varphi_1 - \varphi_2)}{\sin(\varphi_1 + \varphi_2)}. \quad (7\text{-}10)$$

At the second interface, we have

$$r_{2p} = \frac{n_3 \cos \varphi_2 - n_2 \cos \varphi_3}{n_3 \cos \varphi_2 + n_2 \cos \varphi_3} = \frac{\operatorname{tg}(\varphi_2 - \varphi_3)}{\operatorname{tg}(\varphi_2 + \varphi_3)}, \quad r_{2s} = \frac{n_2 \cos \varphi_2 - n_3 \cos \varphi_3}{n_2 \cos \varphi_2 + n_3 \cos \varphi_3} = \frac{\sin(\varphi_2 - \varphi_3)}{\sin(\varphi_2 + \varphi_3)}. \quad (7\text{-}11)$$

As can be seen from Figure 7.5, the incident light will undergo multiple reflections and refractions at the two interfaces, and the total reflected beam result from the interference of many reflected beams. Using the theory of multi-beam interference, the total reflection coefficient of the p and s components is obtained:

$$R_p = \frac{r_{1p} + r_{2p}\exp(2i\delta)}{1 + r_{1p} r_{2p} \exp(-2i\delta)}, \quad R_s = \frac{r_{1s} + r_{2s}\exp(-2i\delta)}{1 + r_{1s} r_{2s} \exp(-2i\delta)}. \quad (7\text{-}12)$$

Among these,

$$2\delta = \frac{4\pi}{\lambda} d n_2 \cos\varphi_2. \quad (7\text{-}13)$$

Equation (7-13) is the phase difference between adjacent reflected beams while λ is the wavelength of light in vacuum.

The change in the polarization state of the beam before and after reflection can be characterized by the total-reflectance ratio (R_p/R_s). In the ellipsometric method, the parameters φ and Δ are used to describe the reflection coefficient ratio, which is defined as:

$$\text{tg}\,\varphi\exp(\text{i}\Delta)=\frac{R_p}{R_s}. \qquad (7\text{-}14)$$

By analyzing the above equations under conditions determined by λ, φ_1, n_1 and n_3, φ and Δ are only functions of the film thickness d and the refractive index n_2; as long as φ and Δ are measured, in principle, d and n_2 should be solved. However, the specific forms of $d=(\varphi, \Delta)$ and $n_2=(\varphi, \Delta)$ cannot be resolved from the above equations. Therefore, it is only possible to first calculate the relationship between (φ, Δ) and (d, n) under the conditions λ, φ_1, n_1, and n_3 according to the above equations, and to determine the φ and Δ of a certain film. Then, the corresponding d and n (n_2) values can be found from the chart.

The methods of measuring φ and Δ are mainly photometry and extinction. The basic principle of determining φ and Δ by ellipsometry is described below. Let the p and s components of the incident and reflected beam's electrical vectors be E_{ip}, E_{is}, E_{rp}, and E_{rs}, respectively:

$$R_p=\frac{E_{rp}}{E_{ip}},\ R_s=\frac{E_{rs}}{E_{is}}. \qquad (7\text{-}15)$$

Therefore:

$$\text{tg}\,\varphi\exp(\text{i}\Delta)=\frac{E_{rp}/E_{ip}}{E_{rs}/E_{is}}. \qquad (7\text{-}16)$$

To make φ and Δ into physical quantities that can be relatively easily measured, we should try to satisfy the following conditions:

(1) Satisfying the incident beam $|E_{ip}|=|E_{is}|$;

(2) Making the reflected beam into linearly polarized light; that is, making the phase difference between the two components of the reflected light 0 or π.

When the above two conditions are met, we have:

$$\begin{cases}\text{tg}\,\varphi=\pm|E_{rp}|/|E_{rs}|\\ \Delta=(\beta_{rp}-\beta_{rs})-(\beta_{ip}-\beta_{is}).\\ (\beta_{rp}-\beta_{rs})=0\ \text{or}\ \pi\end{cases} \qquad (7\text{-}17)$$

Here, β_{ip}, β_{is}, β_{rp}, β_{rs} are the phases of the p and s components of the incident and reflected beams, respectively.

Figure 7.6 is a schematic illustration of a measurement device. In the figure's coordinate system, the x- and x'-axes are both in the incident plane and perpendicular to the propagation directions of the incident or reflected beams, respectively; the y-axis and y'-axis are perpendicular to the incident surface. The angles of the transmission axes t and t' of the polarizer and the analyzer with respect to the x-axis and the x'-axis are P and A, respectively.

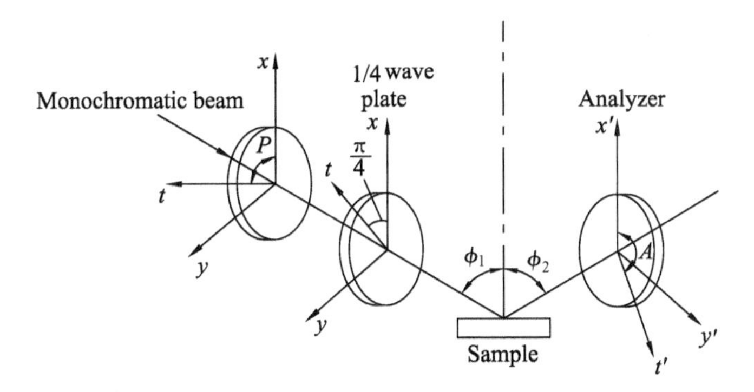

Figure 7.6 Schematic picture of the measurement device[13]

E_0 in Figure 7.7 represents the linearly polarized light emitted by a polarizer with an azimuthal angle of P. When it is projected onto the 1/4-wave plate with the fast axis f and the x-axis at an angle $\pi/4$, it will be decomposed onto the fast axis f and the slow axis s of the wave plate as:

$$E_{f1}=E_0\cos\left(P-\frac{\pi}{4}\right),\quad E_{s1}=E_1\sin\left(P-\frac{\pi}{4}\right). \tag{7-18}$$

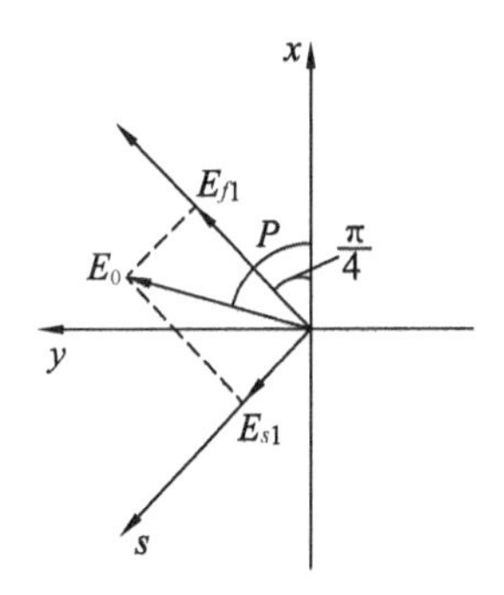

Figure 7.7 Orientation of the fast axis of the 1/4-wave plate

After passing through the 1/4-wave plate, E_f will be ahead of E_s by $\pi/2$, so after the 1/4-wave plate, we should have:

$$E_{f2}=E_0\cos\left(P-\frac{\pi}{4}\right)\exp\left(\mathrm{i}\frac{\pi}{2}\right),\quad E_{s2}=E_{s1}=E_0\sin\left(P-\frac{\pi}{4}\right). \tag{7-19}$$

Projecting these two components onto the x- and y-axes and synthesizing them into E_x and E_y:

$$E_x=E_{f2}\cos\left(\frac{\pi}{4}\right)-E_{s2}\sin\left(\frac{\pi}{4}\right)=\frac{\sqrt{2}}{2}(E_{f2}-E_{s2})=\frac{\sqrt{2}}{2}E_0\exp\left[\mathrm{i}\left(P+\frac{\pi}{4}\right)\right], \tag{7-20}$$

$$E_y=E_{f2}\cos\left(\frac{\pi}{4}\right)+E_{s2}\sin\left(\frac{\pi}{4}\right)=\frac{\sqrt{2}}{2}E_0\exp\left[\mathrm{i}\left(\frac{3\pi}{4}-P\right)\right]. \tag{7-21}$$

It can be seen that E_x and E_y are the p and s components of the incident beam to be projected onto the surface of the sample to be tested,

$$E_{ip}=E_x=\frac{\sqrt{2}}{2}E_0\exp\left[\mathrm{i}\left(P+\frac{\pi}{4}\right)\right],\ E_{is}=E_y=\frac{\sqrt{2}}{2}E_0\exp\left[\mathrm{i}\left(\frac{3\pi}{4}-P\right)\right]. \tag{7-22}$$

Obviously, the incident beam has become a circularly polarized light that satisfies $|E_{ip}|=|E_{is}|$, and the phase difference between the two components is

$$\beta_{ip}-\beta_{is}=2P-\frac{\pi}{2}. \tag{7-23}$$

From Figure 7.8, it can be seen that when the transmission axis t' of the analyzer is perpendicular to the electric vector E_r of the resultant reflected linearly polarized beam (that is, when the reflected light is extinct after the analyzer) we should have:

$$|E_{rp}|/|E_{is}|=\tan A \tag{7-24}$$

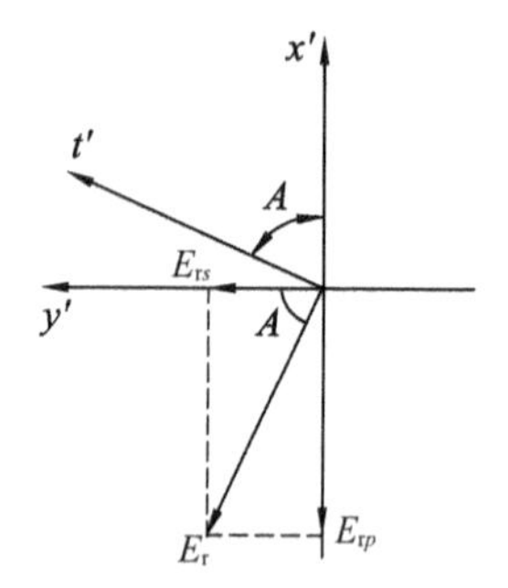

Figure 7.8 Orientation of the transparent axis of the analyzer

Thus, from Formula (7-17), we can obtain that

$$\begin{cases}\mathrm{tg}\ \varphi=\mathrm{tg}\ A\\ \Delta=(\beta_{rp}-\beta_{rs})-(2P-\frac{\pi}{2})\\ (\beta_{rp}-\beta_{rs})=0 \text{ or } \pi.\end{cases} \tag{7-25}$$

It is apparent that A takes values only in the first and fourth quadrants in the coordinate system (x', y'). The following is a discussion of the cases when $(\beta_{rp}-\beta_{rs})=0$ or π.

(1) $(\beta_{rp}-\beta_{rs})=\pi$: here, P is denoted as P_1, the E_r of the synthesized reflected linearly polarized light is in the second or fourth quadrants, and A is in the first quadrant and denoted by A_1. From Equation (7-25),

$$\begin{cases}\varphi=A_1\\ \Delta=\frac{3\pi}{2}-2P_1.\end{cases} \tag{7-26}$$

(2) $(\beta_{rp}-\beta_{rs})=0$; here, P is denoted as P_2 and the synthesized reflected linearly polarized light E_r is in the first and third quadrants, such that A is in the fourth quadrant and is recorded as A_2, which can be obtained by Equation (7-25) as

$$\begin{cases}\varphi=-A_2\\ \Delta=\frac{\pi}{2}-2P_2.\end{cases} \tag{7-27}$$

From Equation (7-26) and (7-27), the relationship between (P_1, A_1) and (P_2, A_2) is:

$$\begin{cases}A_1=-A_2\\ P_1=\frac{\pi}{2}+P_2.\end{cases} \tag{7-28}$$

Therefore, for the device in Figure 7.6, the angle between the fast axis f of the quarter-wave plate and the x axis is $\pi/4$ and then the azimuth of the analyzer (P_1, A_1) or (P_2, A_2), (φ, Δ), can be obtained according to Equation (7-26) or Equation (7-27), thereby completing the measurement of the total reflection-coefficient ratio. Then, using the calculated relationship chart of π and (d, n), the thickness and refractive index of the film to be tested can be found.

Incidentally, when both n_1 and n_2 are real numbers, $d_0 = \frac{\lambda}{2n_2\cos\varphi_2} = \frac{\lambda}{2} \Big/ \sqrt{n_2^2 - n_1^2\sin^2\varphi_1}$ is also a real number. d_0 is called a thickness period, because it can be seen from Equation (7-13) that, when the thickness of the film is increased by d_0, the corresponding phase difference 2δ is also changed by 2π. This will make the films having the thickness of multiple d_0 have the same (φ, Δ) value. The relationship diagrams of (φ, Δ) and (d, n) are given based on the values in the first period. So the thickness period of film should be determined by other methods. However, films generally measured using the ellipsometry method have a thickness in the first cycle, that is, between 0 and d_0. The ability to measure small thicknesses (nanoscale) is an advantage of ellipsometry.

The complex refractive index of the metal can also be measured by ellipsometry. The metal re-exposure rate n_2 can be decomposed into real and imaginary parts, namely:

$$n_2 = N - iNK. \tag{7-29}$$

The coefficients N and K and ellipsometric angles φ and D in the above equation have the following approximate relationship:

$$N \approx (n_1 \sin\varphi_1 \mathrm{tg}\,\varphi_1 \cos 2\varphi)/(1+\sin 2\varphi\cos\Delta),\ N \approx \mathrm{tg}\,2\varphi\sin\Delta. \tag{7-30}$$

It can be seen that, after measuring the ellipsometric parameters φ and D corresponding to the total reflection-coefficient ratio of the metal sample to be tested, an approximate value of the complex refractive index n_2 can be obtained.

7.2 Balance method

7.2.1 Microbalance method

This method measures the film thickness by directly weighing the film deposited on the substrate such that the balance used must meet several requirements. First, the balance should have a sufficiently high sensitivity (usually on the order of 10^{-3}g/m^2); second, it must be mechanically rigid and not easily deformed; third, the system should be easy to degas at higher temperatures and have a non-periodic damping to suit use in an ultra-high vacuum. At present, the relevant measuring systems are almost all made of quartz.

Assuming a uniform film distribution with mass m, density ρ, and film area S, the film thickness can be determined by:

$$d=\frac{m}{\rho S}. \tag{7-31}$$

The advantage of the microbalancing method is the ability to determine the absolute value of the deposition quality. A disadvantage of this method is that the spatial distribution of the film thickness cannot be given, so the average thickness of the film on the area S is obtained. In addition, the actual specific gravity of the film needs to be obtained by other methods, and its value varies greatly depending on the deposition conditions. If the specific gravity of the corresponding block is used instead, since the actual density of the film is generally less than the mass density of the corresponding block, the measured film thickness will be slightly smaller than the actual film thickness.

7.2.2 Quartz-crystal oscillation method

This method is a dynamic weighing method. The film is deposited on a polished quartz electrode with electrodes attached to both the upper and lower sides of the oscillating circuit. The natural oscillation frequency of a quartz crystal is related to its mass. Depositing a thin film on a quartz wafer changes its quality and its natural oscillation frequency. The change in mass can be obtained by measuring that in natural frequency, and the film thickness can then be obtained, as shown in Figure 7.9.

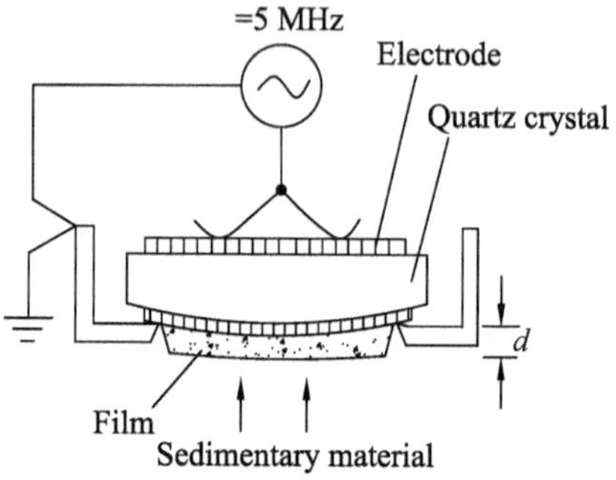

Figure 7.9 Schematic diagram of the probe for a quartz-crystal oscillator[13]

Let the density of the quartz crystal be ρ (2.65 g/cm^3) and the shear elastic modulus be Y; then the velocity of the elastic transverse wave in the thickness t direction of the quartz crystal is:

$$v=\sqrt{\frac{Y}{\rho}}. \tag{7-32}$$

The resonant fundamental frequency of a quartz wafer of thickness t is:

$$f=\frac{v}{\lambda}=\frac{v}{2t}=\frac{N}{t}, \tag{7-33}$$

where $N=\sqrt{\frac{Y}{\rho}}/2$ is the frequency constant and for a quartz crystal, N=1,670 kHz · mm.

Derivation of Formula (7-33) by t is:

$$df=-\frac{N}{t^2}dt=-\frac{f^2}{N}dt. \tag{7-34}$$

It can be seen that the change in thickness is proportional to the change in the oscillation frequency, and the negative sign in the equation indicates that the frequency of the quartz crystal decreases as its thickness increases. If the film thickness (mass-film thickness) deposited on the quartz crystal is d_x, the equivalent quartz-crystal-thickness change is

$$dt=\frac{\rho_d}{\rho}dx, \tag{7-35}$$

where ρ_d is the density of the deposited film.

When $\rho_d\ d_x$ is in a small range, substituting Equation (7-35) into Equation (7-34) yields

$$df=-\frac{f^2}{N}\cdot\frac{\rho_d}{\rho}\cdot dx. \tag{7-36}$$

This formula represents the basic formula for the relationship between the change of oscillation frequency and thickness of the film. For a very thin film, the frequency shift DF is small. When the initial frequency of the crystal is f_0 when the film is deposited, then

$$C_{f_0}=\frac{f_0^{\ 2}}{N\rho}. \tag{7-37}$$

This represents the mass-measurement sensitivity of the system, and its physical meaning is to change the frequency by 1 Hz for the mass increment dm/S, that is, $p_d d_x$.

The corresponding film thickness is

$$dx=-\frac{N}{f^2}\cdot\frac{\rho}{\rho_d}\cdot df. \tag{7-38}$$

It should be noted that the linear relationship between the film thickness dx and frequency shift df given by Equation (7-38) is only established when the df is not large. If the film causes a large frequency shift of the quartz crystal, the f^2 term in Equation (7-38) should be replaced by $(f_0-df)^2\approx f_0(f_0-2df)$.

To satisfy the linear relationship of Equation (7-38), there is a corresponding maximum film thickness that can be measured. From Equation (7-38), it is known that reducing the initial oscillation frequency of the crystal can increase the maximum film thickness that can be measured. However, according to Equation (7-37), using a higher initial frequency can increase the system's sensitivity of the system. In practice, a trade-off must be made between the required sensitivity and the maximum film thickness that can be measured.

It should be noted that the fundamental frequency of the quartz crystal will vary appreciably with temperature, such that the operating temperature of the quartz-wafer probe should not be too high. In addition, since the mechanical properties of the film deposited on the quartz crystal are different from those of the quartz crystal, the effective area and the area of the quartz wafer may be different, and thus Formula (7-35) is only approximate and should be corrected by experiments.

The advantage of this quartz-crystal-oscillation method for measuring the film thickness is that the film thickness can be measured dynamically and continuously during the film-formation process. Moreover, the change in film thickness can be displayed by the frequency. Therefore, if a time-differential circuit is introduced at the output, the growth rate of the film can be measured. The disadvantage is that the measured value is the thickness of the film deposited on the quartz-crystal oscillating plate, which must be recalibrated if the wafer position is changed. In addition, it is susceptible to interference from RF signals when using RF magnetron-sputtering devices.

Quartz-crystal film-thickness monitors can be used on resistive or electron-beam-evaporation equipment to monitor the thickness of metal, semiconductor, and dielectric films. The highest sensitivity of this method is that the frequency change is about 20 Hz, and the mass-film thickness converted into quartz crystal is 12 Å; that is, the accuracy of the measurable film-thickness value is on the order of nm.

A common disadvantage of all methods for measuring film thickness using the balance method is that the density of the film is known. Obviously, the density of the film varies with the process conditions for preparing it, and the relevant error analysis must be considered.

7.3 Electrical method

7.3.1 Resistance method

The resistance method is the simplest means of measuring the thickness of a metal film. Since the resistance of the metal conductive film decreases as film thickness increases, the thickness can be monitored by an electric-resistance method.

As the film thickness decreases, the increase rate of the resistance is greater than expected. The reason for this is scattering at the interface of the film and the difference in the structure of the film from that of the block, as well as the influence of the attached and absorbed residual gases on the electrical resistance. Additionally, the ultrathin film is discontinuous, has an island structure, and its characteristics are completely different from those of the continuous film. Nevertheless, resistance measurements are suitable for a relatively wide range of thicknesses, especially at higher deposition rates and low residual gas pressures.

If the resistivity ρ of the film is known, the film thickness d can be determined by the following formula:

$$d=\frac{\rho}{R_\delta}. \tag{7-39}$$

In the formula, R_δ is often referred to as the square or surface resistance, and has units of Ω. The square resistance is a parameter that is often used in practice and can be directly measured using the four-probe method. Knowing the square-resistance value of the film, the film thickness can be obtained according to Formula (7-39). The resistance of the film is also measured by the Wheatstone-bridge method.

Measuring the film thickness by the resistance method also depends on how precisely the relationship between resistivity and thickness is specified. In fact, the accuracy of the thickness measurement of the metal film by the resistance method is rarely better than 5%.

7.3.2 Capacitance method

A dielectric film sandwiched between two pieces of metal forms a capacitor, whose value is related to the film's thickness. Therefore, this thickness can be determined by measuring its capacitance. According to this principle, the interdigital electrode pair can be deposited on the insulating substrate to form a flat interdigital capacitor. As shown in Figure 7.10, when the dielectric is not deposited, the material between the interdigital electrodes is the substrate, so the capacitance of the interdigital capacitor is mainly determined by the substrate's dielectric constant. If the dielectric film is deposited between the interdigital electrodes, the capacitance is determined by the distance between them, the thickness of the deposited film, and the dielectric constant. If this constant is known, the film thickness can be determined by measuring the capacitance value with a capacitor

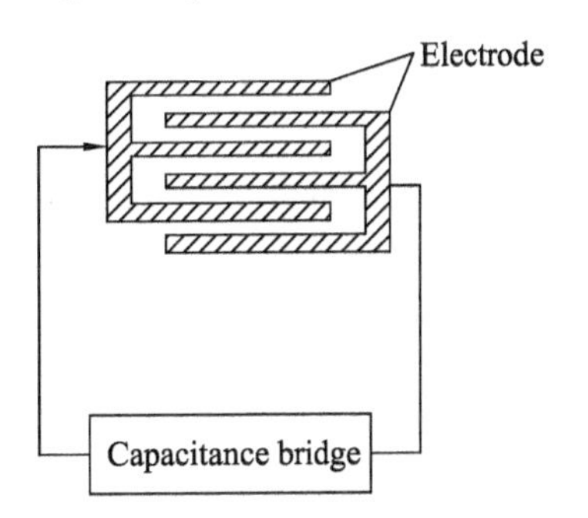

Figure 7.10 Capacitance measurement of film thickness[13]

bridge.

Another method is to form the lower electrode on the insulating substrate and then deposit a layer of dielectric film before forming the upper electrode into a flat capacitor according to the formula for a flat capacitor. After measuring the capacitance, the thickness of the film can be calculated. Obviously this method can only be used to measure the film thickness after deposition, but not to monitor the deposition process.

7.4 Surface-roughness-meter method

A fine diamond pin with a small diameter (~ 1 microns) is pressed against the surface of the film, and an adjustable pressure of 1 ~ 30 mg is added to the pin. When the stylus moves evenly on the film, its thickness will cause the stylus to move up and down. Record the movement of the diamond needle in the vertical direction and draw the contour of the film surface. The vertical displacement can be magnified thousands or even one million times by mechanical, electronic or optical methods, such that the resolution of the vertical displacement is very high (on the order of nm). This method can be used not only to measure the surface roughness and stress, but also the height of the film step and the geometric thickness of the film. It gives the absolute value of the film thickness.

As the displacement is measured, the substrate's surface is needed as a reference point, so that a step must be made between the film and substrate prior to measurement. The instrument for measuring the thin film thickness using a stylus is called a step instrument or surface profilometer, and its accuracy can reach 0.5 nm.

In an actual instrument, there are several ways to electrically amplify the step displacement obtained by moving the contact pin. These are listed as follows.

7.4.1 Differential-transformer method

The principle of amplifying the up-and-down movement displacement of the contact pin via a differential transformer is shown in Figure 7.11 (a). In the figure, the outputs of coils 2 and 3 are connected in the reverse phase. Due to the linkage between the iron core and the contact pin, the up-and-down movement of the contact pin will lead to a change in the magnetic flux generated by the iron core's movement. Coils 2 and 3 will output the differential electrical signal, amplify the signal, and display the value corresponding to the movement distance of the contact pin, i.e. the film thickness.

7.4.2 Impedance amplification

The principle of the impedance-amplification method is shown in Figure 7.11 (b). When the gap of the inductor changes due to the up-and-down movement of the contact pin, the inductive reactance changes accordingly, in turn changing the coil impedance. By amplifying and displaying the impedance variation, the displacement of the up-and-down movement of the contact pin can be characterized.

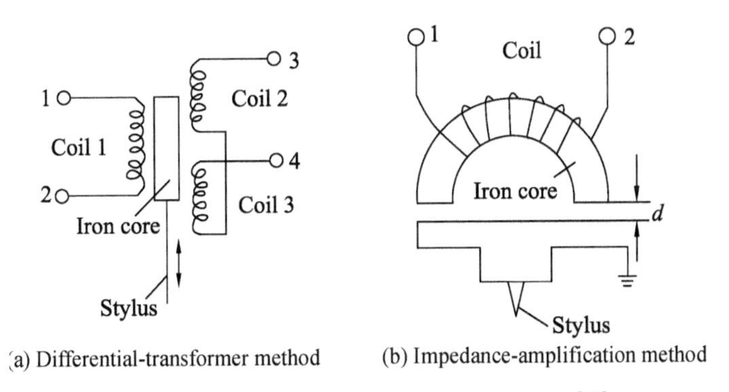

Figure 7.11 Contact-pin thickness sensor[13]

7.4.3 Piezoelectric-element method

The piezoelectric-element method uses the piezoelectric effect of certain materials to enlarge and display the contact pin's moving distance. As the pin moves up and down, the pressure acting on the piezoelectric-crystal element will change accordingly, resulting in changes to the element's electrical parameters. By enlarging and displaying the variation of the electrical parameters, the displacement value of the up-and-down movement of the contact pin can be characterized.

With its advantages of a simple method, intuitive measurement, and guaranteed accuracy, the stylus method has been widely used in the measurement of hard film thickness. However, the following points should be noted:

(1) It is easy to scratch the soft film and cause measurement error. Because the area of the contact pin's tip is very small, it will make a groove on the soft plasma membrane, causing damage and resulting in large error. Therefore, the load of the contact pin and the diameter of the contact should be adjusted according to the softness and hardness of the membrane.

(2) The "noise" caused by the undulation or unevenness of the film surface and the substrate surface will also incur large errors.

Exercises

1. Please describe the principle of optical interferometry method for thin film thickness measurement.
2. Please describe the principle of ellipsometry method for thin film thickness measurement.
3. Please describe the principle of quartz-crystal oscillation method for thin film thickness measurement.
4. Please describe the principles of resistance method and capacitance method for thin film thickness measurement. What are the advantages of them?
5. What are the advantages and disadvantages of piezoelectric-element method for thin film thickness measurement?

Chapter 8 Recent Advances of Thin Film Materials and Technology

As the application range of thin film materials is wider and wider, their position in high technologies becomes more and more important. This greatly stimulates the research of new thin film materials and their manufacturing technologies. In the area of thin film materials and technology, one research direction is to discover new thin film materials for new applications. The other direction is to improve the quality of existing film materials by designing new equipment using novel methods.

8.1 Recent advances in thin film fabrication technologies

In the sputtering technology mainly using electric field, various magnetron sputtering technologies with high speed, low temperature and low damage have been developed by introducing magnetic field. Active reactive gas is introduced into the film-forming process such as evaporation, sputtering and ion plating. Reactive evaporation, reactive sputtering and reactive ion plating have emerged. Thus, the boundary between physical process and chemical process is broken. *The PVD technology is combined with CVD technology to develop new types of thin film materials.* Some of the progresses are listed below:

(1) In order to promote the effectivity of film-forming process with the above mentioned physical-chemical processes, various energy sources (such as laser, ultraviolet light, high-frequency microwaves, ion beam and ionizing electric field, etc.) are introduced into the film-forming process.

(2) In order to produce multi-component or multi-layer composite films, multi-source co-evaporation (or sequential evaporation), multi-target co-sputtering (or sequential sputtering), multi-target ion beam co-sputtering have been developed. Some specially designed composite sputtering targets have also been developed.

(3) In addition to the traditional sol-gel method, some new chemical solution methods combined with spin-coating method have also been developed to fabricate high quality thin films with low cost. A typical example of them is polymer-assisted deposition (PAD) or polymer-assisted metal deposition method (PAMD).[20,21] PAMD uses polymer to control the viscosity and binds metal ions, resulting in a homogeneous distribution of metal precursors in the solution and the formation of uniform metal-organic films. The latter feature makes it possible to grow simple and complex crack-free epitaxial metal-oxides. It also has merits to make thin films on substrates with various shapes, as shown in Figure 8.1. PAMD has received great success in fabricating flexible, stretchable, and even washable metal conductors in the past decade, which are applied in a wide range of flexible devices in sensors, solar cells, transistors, super-capacitors, and batteries.

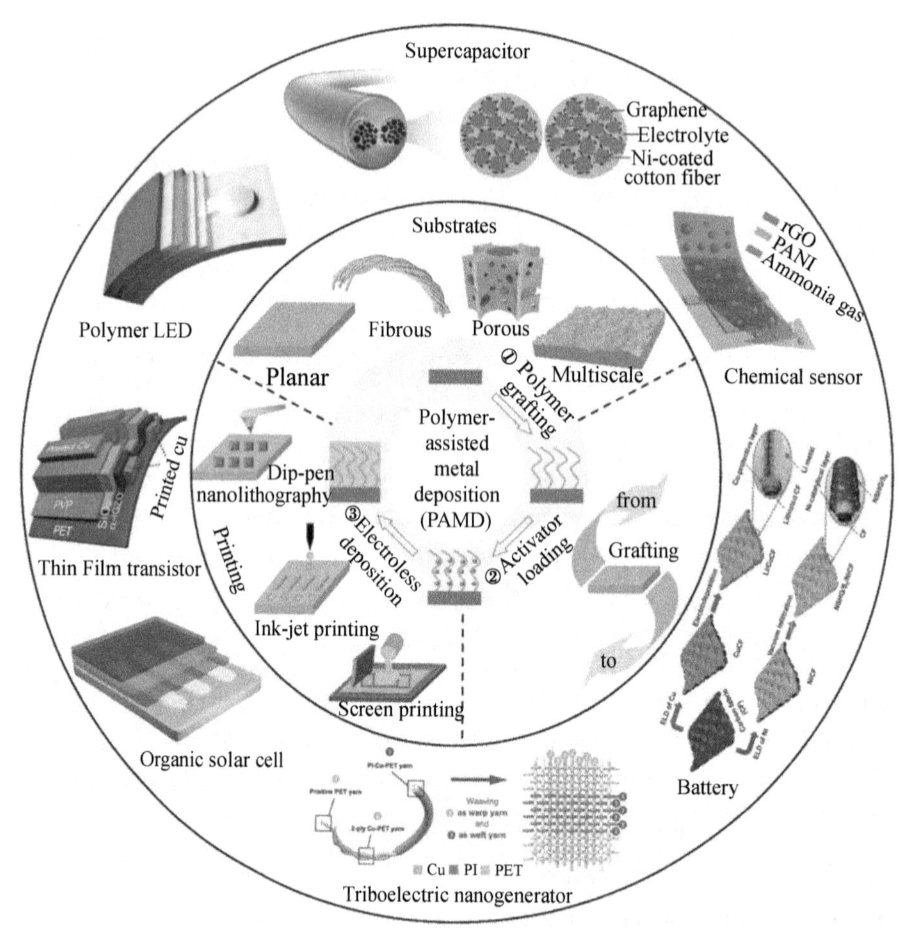

Figure 8.1 Schematic illustration of the mechanism, materials, chemistry, printing compatibility, and applications of polymer-assisted deposition[21]

Due to the adoption of these new technologies, the physical properties of various traditional films are improved. A variety of new functional films are also discovered, such as high temperature(T_c) superconducting films, diamond films, ultrafine particle films, multicomponent lead oxide films and magneto-optical films. They will be introduced in the following contents.

8.2 Some important electronic thin film materials

8.2.1 Classifications of thin film materials

Currently, there are many kinds of electronic thin film materials with wide applications. If classified by categories, they include: superconducting film, conductive film, resistance film, semiconductor film, dielectric film, insulation film, passivation and protection film, piezoelectric film, ferroelectric film, pyroelectric film, photoelectric film, magnetic film, magneto-optical film, etc. Some of the films have single physical properties, while others have two or more excellent properties and thus multi-applications. They can be used for several different purposes. Some of the most-commonly used electronic thin film materials are listed in Table 8.1.

Table 8.1 Some of the most-commonly used electronic thin film materials[2]

Types	Film materials
Superconducting thin film	Nb, Nbn, MoN, Nb_3Sn, Nb_3Ge, Nb_3SiNb_3C, V_3C, NbC_xN_{1-x}; $PbMo_6S_8$, $Li_{1-x}Ti_{2-x}O_4$, $M_xLa_{2-x}CuO_{4-x}$ (M = Ba, Sr, Ca), $BaPb_{1-x}Bi_xO_3$; YBaCuO-series, BiSrCaCuO-series, TiBaCaO-series
Conductive thin film	Au, Al, Cu, Cr, Ni, Ti, Pt, Pd, Mo, W; Al-Si, Pt-Si, Mo-Si, Cr-Au, NiCr-Au, Cr-Cu-Au, Ti-Mo-Au; ZnO, In_2O_3, SnO_2, TiO_2, Cd_2SnO_4
Resistance thin film	Cr, Ta, Re, NiCr, CrSi, CrTi, TaAl, TaSi, ZrB_2; TaN, TiN, TaAlN, SnO_2, In_2O_3; Cr-SiO, Cr-SiO_2, Au-SiO, Au-SiO_2, Ta-Al_2O_3
Semiconductor thin film	Ge, Si, Se, Te, SiC, GaAs, GaP, GaN; ZnO, ZnSe, ZnTe, ZnCdS, CdSe, CdTe, CdS, PbS, PbO_2, Mn-Co-NiO; a-Si: H, As_2S_3, As_2Se_3, As_2Te_3, GeTe
Dielectric thin film	BN, AlN, Si_3N_4, ZnS; BeO, Al_2O_3, SiO, SiO_2, TiO_2, HfO_2, ZrO_2, PbO, MgO, Y_2O_3, Ta_2O_2, Nb_2O_5; $BaTiO_3$, $PbTiO_3$, $LiNbO_3$, PLZT
Insulating thin film	Si_3N_4, SiO, SiO_2, Al_2O_3, TiO_2, Ta_2O_5, PIQ
Protecive thin film	Si_3N_4, SiO, SiO_2, PSG, BPSQ, PIQ, AlN, Diamond film
Ferroelectric thin film	$BaTiO_3$, $PbTiO_3$, $Pb(Zr_{1-x}Ti_x)O_3$ (PZT), PLZT ($PbTiO_3$-$PbZrO_3$-La_2O_3), $BiFeO_3$

continued

Types	Film materials
Piezoelectric thin film	ZnS, ZnO, AlN, $LiNbO_3$, PZT, ZnO-AlN, SiO_2-$LiNbO_3$
Pyroelectric thin film	TGS, $LiTaO_3$, $PbTiO_3$, PZT, PLZT, PVF_2, PVDF; $Sr_xBa_{1-x}Nb_2O_6$
Photoelectric thin film	Si, InP, GaAs, CdTe, CdS, $Al_xGa_{1-x}As$, a-Si:H, a-SiGe, a-SiSn, $BiFeO_3$
Magnetic thin film	$R_3Fe_5O_{12}$ (R = Rare earth metals), MFe_2O_4 (M = Bivalent metal elements); γ-Fe_2O_3, Fe_3O_4; NiCo, NiFe, FePt, CoCrPt, NiMn, FeCo
Magneto-electric thin film	Ge, Si, InSb, InAs, $BiFeO_3$
Magneto optic thin film	$TbFeO_3$, $Y_3Ga_{1.1}Fe_{3.9}O_{12}$, MnBi, PtCo, MnCuBi; Gd-Tb-Fe, Gd-Tb-Co, Tb-Fe-Co, Tb-Co, Gb-Co, Gb-Fe

The following part introduces a few of these thin film materials in more details.

8.2.2 High T_c superconducting thin films

Since the discovery of high-temperature (high-T_c) superconductors in 1896, there has been a continuous worldwide interests in the development of high-T_c superconducting films. Almost all kinds of advanced film-making methods have been tried, and a variety of high-T_c superconducting films have been developed recently. At present, there are three kinds of films with mature technology and good physical properties:

(1) YBaCuO-series film. T_c = 90 K, $J_c = 1\times10^6$ A/cm^2;

(2) BiSrCaCuO-series film. T_c = 110 K, $J_c = 1\times10^5$ A/cm^2;

(3) TiBaCaCuO-series film. T_c = 120 K, $J_c = 1\times10^6$ A/cm^2.

Although the critical temperatures of these films are much higher than that of liquid nitrogen, some problems must be solved to make practical superconducting electronic devices and realize superconducting microelectronics.

(1) First of all, we must develop the ability to grow high T_c superconducting thin film on large area substrates. The requirements of substrate materials for superconducting thin films are as follows: the lattice constant is close to that of superconductor, low temperature stability, good crystallinity, and no reaction with superconductor. At present, MgO and $SrTiO_3$ substrates (90%) are mainly used, and ZrO_2 substrates are also used in China. It has been reported that $SrTiO_3$ substrates have been replaced by new substrate materials, such as $LaAlO_3$ and $LaGaO_3$, etc. Among them, $LaAlO_3$ has been successfully developed in China, but large size single crystal, $LaGaO_3$, cannot be obtained. $LaGaO_3$ is considered to be the most promising substrate material because of its double crystal structure. Komatsu has successfully manufactured and obtained 30 mm × 60 mm high

quality single crystal.

(2) The film-forming temperature should be as low as possible (below 600 ℃ ~ 400 ℃), and it has good process consistency and repeatability.

(3) The long-term stability and compatibility with semiconductor technology of superconducting thin films are studied.

(4) The contact and interconnection with other materials, the influence on superconductivity and the interface characteristics are studied.

(5) The growth technology of superconducting thin films on ultrathin insulating films is studied.

With the solution of the above problems, high-T_c superconducting thin films will be widely used, and the appearance of superconducting electronics will change significantly. Moreover, it is expected that high-T_c superconducting thin films will be used in various microwave passive circuits based on high-T_c superconducting thin film microstrip lines.

8.2.3 Diamond thin films

Diamond thin films has many excellent characteristics.

(1) At room temperature, its thermal conductivity is very fast, which is 6 times of that of copper and almost one order of magnitude higher than that of BeO.

(2) Its transparency is high, and it can pass through various wavelengths of light from ultraviolet to infrared; its sound speed is fast, which is 15~16.5 km/s, 1.7 times of that of titanium based material.

(3) Its resistivity is large, which is $10^6 \sim 10^{12}$ Ω · cm, and can be used as a semiconductor or even a semiconductor insulator. Its band gap width is 5.5 eV, much higher than SiC (2.8 eV).

(4) It is a crystal with the same structure as Si, Ge and other semiconductors, with excellent chemical corrosion resistance. It has strong radiation resistance, especially suitable for military and other harsh application environment.

(5) In addition, the electronic components made of diamond film can keep a low temperature when working, so it will be an ideal component of high-speed electronic computer.

At present, the development of diamond film has become one of the key points of high-tech competition in some industrial developed countries. It is predicted that diamond films will become a new material for the next generation of electronic components. Therefore, in addition to the high research interest of high temperature superconductor film, research interests on diamond film appear in the world.

In order to obtain high-quality diamond films, almost all kinds of methods have been used, such as flame spraying, hot filament CVD, ion beam sputtering, ion beam evaporation and microwave plasma CVD, etc. Among them, hot filament CVD and microwave plasma CVD are the major methods. At present, the main topics in the research of diamond films are:

(1) The low temperature, rapid growth and large-area fabrication technology of diamond films;

(2) The fabrication technology of diamond semiconductors, mainly the doping technology for n-type diamond semiconductor films;

(3) Continuous growth of diamond films on heterogeneous and insulating substrates; the growth mechanism and film-forming technology of dense diamond film;

(4) The application of dense diamond film as thermal lining material in microwave semiconductor devices and lasers, as well as in thermal and photosensitive devices.

With the solution of the above problems, it is expected that diamond films will become a new material in temperature resistance, laser resistance, corrosion resistance, radiation resistance, far infrared, and ultra-violet (UV) light sensitive devices.

8.2.4 Langmuir-Blodgett films

Langmuir-Blodgett (LB) films were first produced in the 1930s. Since the 1970s, due to the improvement of film-forming technology, high-quality LB films with uniform thickness and controllable thickness have been produced. They have a good prospect in practical application and has attracted more and more attention in both research and industry area.

A LB film is a kind of monolayer or multilayered film composed of organic molecules arranged in fixed intervals. Its preparation method is quite different from the previous film-forming technology. It is based on organic polymer materials with hydrophilic and hydrophobic groups as raw materials. Due to the "amphiphilic balance" state, such organic molecules will be adsorbed at the water gas interface. If the surface active substance is dissolved in benzene, dichloromethane and other volatile solvent. The liquid is distributed on the water surface. After the solvent volatilization, the directional molecules standing vertically on the water surface. Then these molecules are connected with each other through special film-forming equipment, and then they are gradually stacked on the surface of solid substrate to form two-dimensional crystal, i.e. a LB film, as shown in Figure 8.2.

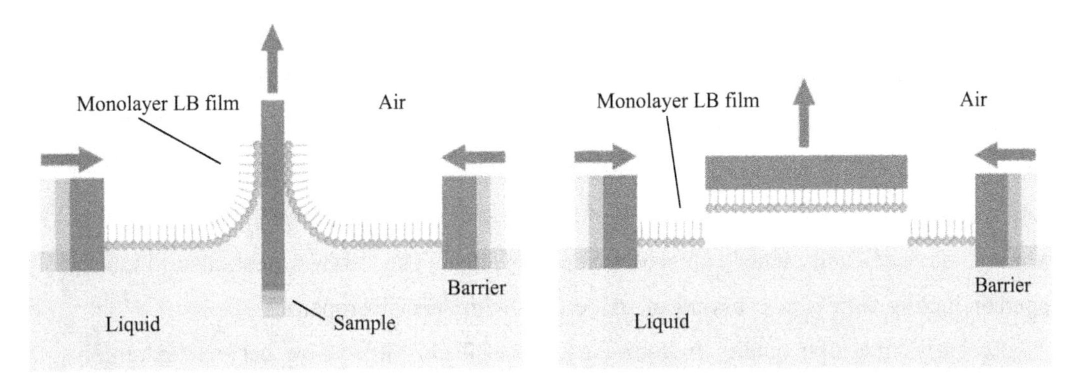

Figure 8.2 Schematic picture of the growth process of monolayer Langmuir-Blodgett (LB) film

According to the needs and adjusting the hydrophilicity and hydrophobicity of raw material molecules, the LB films transferred to the substrate can be single molecular layer or multi-layer layer, can be the same molecular layer or multi-layer structure composed of different molecules.

At present, good results have been obtained in the field effect devices, electroluminescence, integrated optical path and biosensor of MIS structure. In addition, since it is a technology for designing and fabricating new thin film materials at the molecular level, it will undoubtedly be widely used in molecular engineering in the future. Some companies have successfully developed high functional separation membranes and ultra thin films by using this technology. Japan's Asahi Glass Company, for example, has spent three years to find applications of LB films in medicine industry. LB films can filter blood taken from the body and remove toxic substances. It is suggested that the human body is composed of a membrane which can separate necessary and unnecessary substances at normal temperature and pressure. Therefore, scholars in Japan and the United States put forward the prediction that "the LB film maker will make the chemical industry of tomorrow" and "the LB film maker will control the world".

8.2.5 Oxide films

In recent years, the film forming of functional ceramics or electronic ceramics with excellent electrical, magneto-optical and mechanical properties has attracted great attention. This is not only due to the emergence of oxide high-T_c superconducting materials in 1896 which used the traditional ceramic technology, but also to meet the increasing demands on small and light-weight electronic devices.

PVD, CVD and the combination of PVD and CVD are still the main methods of thin film technology of oxide materials. In addition, Sol-gel method has gained a new life on this turning point, which has been highly valued by film making industries due to its low cost.

At present, ZnO thin films have been applied in surface acoustic wave devices and optical waveguides. In addition, active research has been carried out in $PbTiO_3$, $Pb(Zr_{1-x}Ti_x)O_3$ (PZT) and PLZT ($PbTiO_3$-$PbZrO_3$-La_2O_3) films. It is expected that these films will be widely used in pyroelectric devices, optical modulators, optical memory devices and other optoelectronic devices. The above-mentioned high-T_c superconducting film is also a kind of oxide film with special properties.

Recently, the high quality Indium-Tin oxides (ITO) films have become extremely important in both scientific research and industry due to their high transparency and conductivity. Liquid-crystal displays, the multitude of organic light-emitting diode (OLED) variants, and most other flat-panel display technologies rely on transparent conductive oxides (TCOs) such as indium tin oxide (ITO) to transport current and to serve as the anode for each light-emitting element, as shown in Figure 8.3. It can also be used as the top electrodes in flexible solar cell devices.

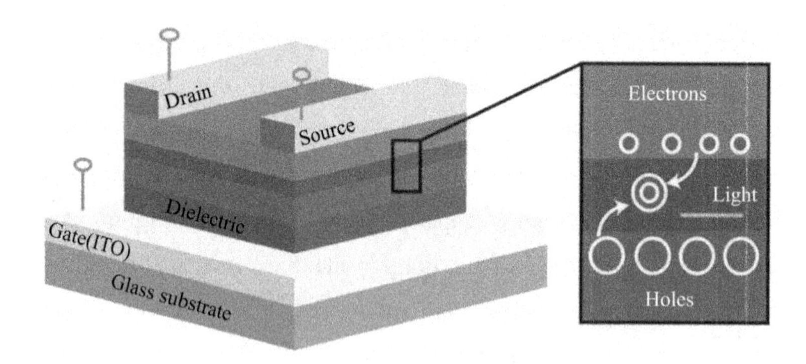

Figure 8.3 Schematic picture of a LED device with transparent ITO film as gate layer

8.2.6 Super-lattice films

Super-lattice thin films are multilayer films with two or more different film components growing alternately in thin layers and maintaining its strict periodicity. The thickness of each sublayer in super-lattice multilayer thin films is usually less than the mean-free-path of electrons in the material, but larger than the lattice constant of the material. Generally, the thickness of each layer in super-lattice films range from one to tens of nanometers.

Since the development of super-lattice films in 1970s, this kind of thin films has developed rapidly due to the development of advanced thin film growth techniques such as molecular beam epitaxy technology. The main methods of fabricating superlattice films are molecular beam epitaxy, ion beam sputtering and pulsed laser deposition devices.

By changing the composition and thickness of the sublayer, new materials can be designed and manufactured artificially in the range of atomic linearity. It is especially suitable for making high quality compound semiconductor mixed crystal materials. Such as CaAlAs/GaAs, HgTe/CdTe, a-Si: H/a-SiC_x, etc. It will be widely used in the development of semiconductor lasers, infrared detectors, photoelectric devices and so on. Recently, super-lattice oxide thin films have also been widely studied and some novel physical properties are discovered in these super-lattice films. A typical example of these is the discovery of high-T_c super-conductivity at FeSe/$SrTiO_3$ interfaces by the research group of Prof. Qikun Xue.[22] Recently, the synthetic antiferromagnets has also been discovered in super-lattice oxide multilayers,[23] as shown in Figure 8.4.

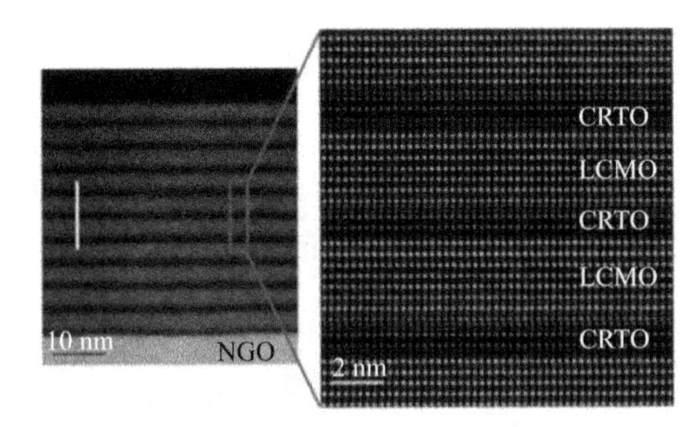

Figure 8.4 TEM images of local atomic structure of [LCMO/CRTO]$_n$ synthetic antiferromagnets super-lattice multilayer oxide thin films[23]

 8.3 Progresses of two-dimensional thin film materials

Two dimensional (2D) materials are a general term for a special class of thin film materials whose thickness is atomically thin. Graphene is the earliest discovered and most typical 2D material which is atomically thin carbon films, as shown in Figure 8.5. In 2004, K. S. Novoselov et al, published an article in *Science* magazine, reporting that single-layer graphene was obtained by mechanical exfoliation of graphite.[24] The unique and excellent electrical properties of graphene were found. This finding led to a Nobel prize in physics at 2010.

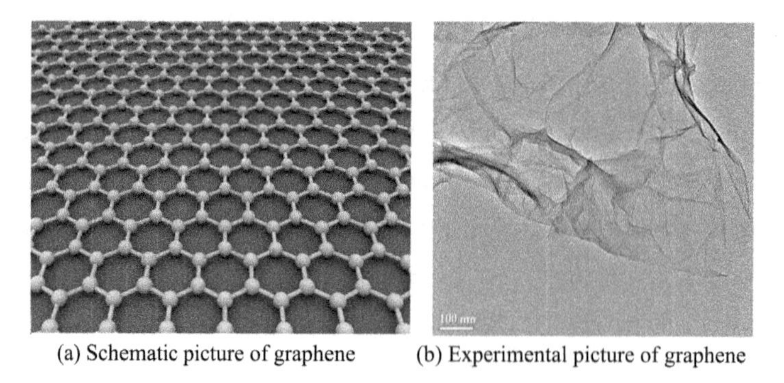
(a) Schematic picture of graphene (b) Experimental picture of graphene

Figure 8.5 Schematic picture of graphene and experimental picture of graphene

Since graphene was discovered, the research work on 2D materials have been developed rapidly, and many new two-dimensional materials were discovered. Other 2D materials include: single-element silene, germanene, stannene, bornene and black phosphorus; transition metal chalcogenides such as MoS_2 (Figure 8.6), WSe_2, ReS_2, $PtSe_2$, $NbSe_2$, etc.; main group metal chalcogenide compounds such as GaS, InSe, SnS, SnS_2, etc.; and other 2D materials such as h-BN, CrI_3, $NiPS_3$, Bi_2O_2Se, etc.[26,27]

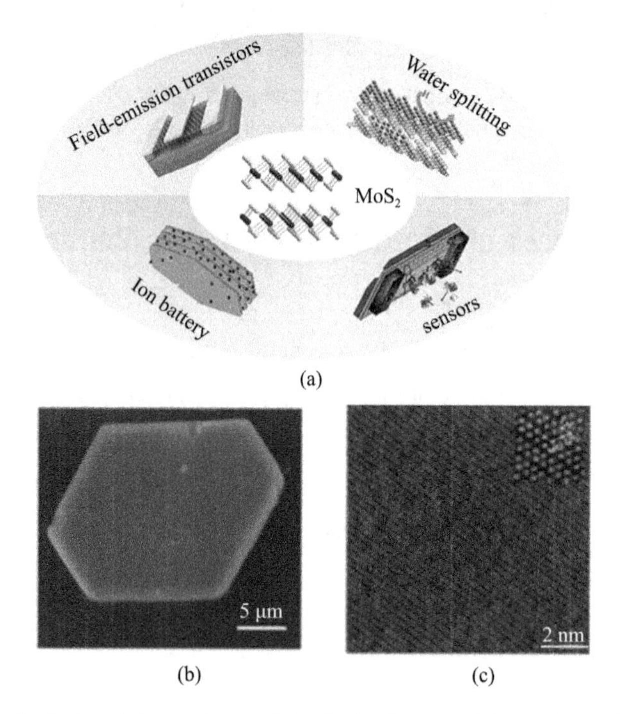

(a)

(b) (c)

Figure 8.6 (a) Crystal structure of typical 2D material MoS_2 and its potential applications, (b) the SEM and (c) TEM images of a MoS_2 thin film sample[25]

Due to the quantum confinement effect in the thickness direction of the atomic layer,

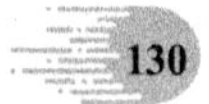

these 2D materials show different properties from their corresponding three-dimensional structures. They have different band structures and electrical properties, covering materials from superconductors, metals, semi metals, semiconductors to insulators. At the same time, they also have special optical, mechanical, thermal and magnetic properties. By stacking different kinds of 2D materials, a more functional material system can be constructed. Therefore, these materials are expected to be used in high-performance electronic devices, optoelectronic devices, spin electronic devices, energy conversion and storage devices. Although it is rather hard to replace silicon materials, 2D materials can complement the existing technical functions of devices.

The research of 2D materials is facing many challenges. First of all, the preparation level of the materials is far from the standard of optoelectronic devices. Although most of the 2D layered materials can be obtained by mechanical exfoliation, the efficiency of this method is low, the transverse size of the sample is small, and the thickness is not easy to control. Large area graphene and some 2D materials can be prepared by other methods such as liquid phase exfoliation or chemical vapor deposition. However, the parameters such as the number of layers, edge morphology, defect density, phase and impurity concentration of the samples are difficult to control, and these methods need to be further optimized in the preparation of new 2D materials. Therefore, in order to realize the wide application of 2D materials, controllable preparation is the preferable, which involves the advanced thin film technologies for the preparation of large area 2D materials for industrial applications.

Exercises

1. Please list some examples of the recent advances of thin film fabrication technologies.
2. Among the most commonly used electronic thin film materials, which materials have been used in the microelectronic transistor technologies? Please list some examples.
3. Among the most commonly used electronic thin film materials, which materials have been used in the renewable energy technologies? Please list some examples.
4. What are the definition and special properties of super-lattice thin films?
5. What is the definition of two-dimensional thin film material? Please list some examples of two-dimensional thin film materials.

References

［1］ 田民波,刘德令．薄膜科学与技术手册［M］.北京:机械工业出版社,1991.

［2］ 杨邦朝,王文生．薄膜物理与技术［M］.成都:电子科技大学出版社,1994.

［3］ 唐伟忠．薄膜材料的制备原理、技术及应用［M］.北京:冶金工业出版社,1998.

［4］ 李学丹,万英超,姜祥祺,等．真空沉积技术［M］.杭州:浙江大学出版社,1994.

［5］ 王力衡,黄运添,郑海涛．薄膜技术［M］.北京:清华大学出版社,1992.

［6］ 薛增泉,吴全德,李浩．薄膜物理［M］.北京:电子工业出版社,1991.

［7］ 曲喜新．薄膜物理［M］.上海:上海科学技术出版社,1986.

［8］ 田民波．薄膜技术与薄膜材料［M］.北京:清华大学出版社,2006.

［9］ 王欲知．真空技术［M］.北京:机械工业出版社,1987.

［10］ 赵宝升．真空技术［M］.北京:科学出版社,1998.

［11］ 张光华．等离子体与成膜基础［M］.北京:国防工业出版社,1994.

［12］ 郑伟涛,等．薄膜材料与薄膜技术［M］.北京:化学工业出版社,2004.

［13］ 宁兆元,江美福,辛煜,等．固体薄膜材料与制备技术［M］.北京:科学出版社, 2008.

［14］ 王建祺,吴文辉,冯大明．电子能谱学(XPS/XAES/UPS)引论［M］.北京:国防工业出版社,1992.

［15］ CARTER C B, NORTON M G. Ceramic materials science and engineering［M］. New York: Springer, 2007.

［16］ SMITH D L. Thin-film deposition: principles and practice［M］. New York: McGraw-Hill, Inc., 1995.

［17］ HIPPEL A V. Dielectrics and waves［M］. London: Artech House Press, 1995.

［18］ OHRING M. Materials science of thin films［M］. New York: Academic Press,2002.

［19］ GESSERT T. Vacuum pumps: an introduction to their operation and maintenance［EB/OL］.［2020－11－01］. https://avs.org/education-outreach/short-courses/short-course-catalog/vacuum-equipment-technology/system-pump-operation/vacuum-pumps-an-introduction-to-their-operation/.

［20］ JIA Q X, MCCLESKEY T M, BURRELL A K, LIN Y, et al. Polymer-assisted deposition of metal-oxide films[J]. Nature materials, 2004, 3: 529-532.

[21] LI P, ZHANG Y K, ZHENG Z J. Polymer-Assisted Metal Deposition (PAMD) for flexible and wearable electronics: principle, materials, printing, and devices[J]. Advanced materials, 2019, 31: e1902987.

[22] WANG Q Y, LI Z, ZHANG W H, et al. Interface-induced high-temperature superconductivity in single unit-cell FeSe films on $SrTiO_3$[J]. Chinese physics letters, 2012, 29: 037402.

[23] CHEN B B, XU H R, MA C, et al. All-oxide-based synthetic antiferromagnets exhibiting layer-resolved magnetization reversal[J]. Science, 2017, 357: 191-194.

[24] NOVOSELOV K S, GEIM A K, MOROZOV S V, et al. Electric field effect in atomically thin carbon films [J]. Science, 2004, 306: 666-669.

[25] X-MOL. 1T‴-MoS_2 晶体的结构解析及其非线性光学研究[EB/OL]. (2019-04-17) [2021-04-26]. https://www.x-mol.com/news/17036.

[26] FANG Y Q, HU X Z, ZHAO W, et al. Structural determination and nonlinear optical properties of new 1T‴-type MoS_2 compound [J]. Journal of the American Chemical Society. 2019, 141(2): 790-793.

[27] WANG H, YUAN H, HONG S S, et al. Physical and chemical tuning of two-dimentional transition metal dichalcogenides[J]. Chemical Society reviews, 2015, 44: 2664 -2680.